LIKE A THIEF IN BROAD DAYLIGHT

LIKE A THIEF IN BROAD DAYLIGHT

POWER IN THE ERA OF POST-HUMAN CAPITALISM

SLAVOJ ŽIŽEK

Seven Stories Press
New York • Oakland • London

Originally published in the UK by Allen Lane, 2018.
First US edition October 2019.

Seven Stories Press
140 Watts Street
New York, NY 10013
www.sevenstories.com

College professors and high school and middle school teachers may order free examination copies of Seven Stories Press titles. To order, visit www.sevenstories.com, or fax on school letterhead to (212) 226-1411.

Library of Congress Cataloging-in-Publication Data has been applied for.

ISBN: 978-1-60980-975-1 (pbk)
ISBN: 978-1-60980-976-8 (ebook)

Printed in the USA.

9 8 7 6 5 4 3 2 1

To Jela, with l—!

Contents

Introduction: First the Bad News, Then the Good News . . . Which May Be Even Worse

Alain Badiou's *The True Life*[1] opens with the provocative claim that, from Socrates onwards, the function of philosophy is to corrupt the youth, to alienate (or, rather, 'extraneate' in the sense of Brecht's *verfremden*) them from the predominant ideologico-political order, to sow radical doubts and enable them to think autonomously. The young undergo the educational process in order to be integrated into the hegemonic social order, which is why their education plays a pivotal role in the reproduction of the ruling ideology. No wonder that Socrates, the 'first philosopher', was also its first victim, ordered by the democratic court of Athens to drink poison. And is this prodding not another name for evil – evil in the sense of disturbing the established way of life? All philosophers prodded: Plato submitted ancient customs and myths to ruthless rational examination, Descartes undermined the medieval harmonious universe, Spinoza ended up being excommunicated, Hegel unleashed the all-destructive power of negativity, Nietzsche demystified the very basis of our morality . . . even if they sometimes appeared almost as state-philosophers, the establishment was never really at ease with them. We should also consider their counterparts, the 'normalizing' philosophers who tried to restore the lost balance and reconcile philosophy with the established order: Aristotle with regard to Plato, Thomas Aquinas with regard to effervescent early Christianity, post-Leibnizian rational theology with regard to Cartesianism, neo-Kantianism with regard to post-Hegelian chaos . . .

Is the pairing of Jürgen Habermas and Peter Sloterdijk not the latest incarnation of this tension between prodding and normalization, shown in their reaction to the shattering impact of modern sciences, especially brain sciences and biogenetics? The progress of today's sciences destroys the basic presuppositions of our everyday notion of reality.

There are four main attitudes one can adopt towards this breakthrough. The first one is simply to insist on radical naturalism, i.e. to heroically pursue the logic of the scientific 'disenchantment of reality' whatever the cost, even if the very fundamental coordinates of our horizon of meaningful experience are thereby shattered. (In brain sciences, Patricia and Paul Churchland most radically opt for this attitude.) The second is to make a desperate attempt to move beneath or beyond the scientific approach into some presumably more original and authentic reading of the world (religion or other kinds of spirituality are the main candidates here) – as, ultimately, Heidegger does. The third and most hopeless approach is to try to forge some kind of New Age 'synthesis' between scientific Truth and the premodern world of Meaning: the claim is that new scientific results themselves (quantum physics, say) compel us to abandon materialism and point towards some new (Gnostic or Eastern) spirituality. Here is a standard version of this idea:

> The central event of the twentieth century is the overthrow of matter. In technology, economics, and the politics of nations, wealth in the form of physical resources is steadily declining in value and significance. The powers of mind are everywhere ascendant over the brute force of things.[2]

This line of reasoning stands for ideology at its worst. The reinscription of proper scientific problematics (the role of waves and oscillations in quantum physics, for example) into the ideological field of 'mind versus brute things' obfuscates the truly paradoxical result of the notorious 'disappearance of matter' in modern physics: how the very 'immaterial' processes lose their spiritual character and became a legitimate topic of natural sciences.

None of these three options is adequate for the establishment, which basically wants to have its cake and eat it: it needs science as the foundation of economic productivity, but it simultaneously wants to keep the ethico-political foundations of society free from science. In this way, we arrive at the fourth option, a neo-Kantian state philosophy whose exemplary case today is Habermas (but there are others, like

Luc Ferry in France). It is a rather sad spectacle to see Habermas trying to control the explosive results of biogenetics and curtail its philosophical consequences – his entire endeavour betrays the fear that something will happen, that a new dimension of the 'human' will emerge, that the old image of human dignity and autonomy will survive unscathed. Over-reaction is common here, such as the ridiculous response to Sloterdijk's Elmau speech on biogenetics and Heidegger,[3] discerning the echoes of Nazi eugenics in the (quite reasonable) proposal that biogenetics compels us to formulate new rules of ethics. Techno-scientific progress is perceived as a temptation that can lead us into 'going too far' – entering the forbidden territory of biogenetic manipulations and so on, and thus endangering the very core of our humanity.

The latest ethical 'crisis' apropos biogenetics effectively creates the need for what one is fully justified in calling a 'state philosophy': a philosophy that would, on the one hand, promote scientific research and technical progress and, on the other, contain its full socio-symbolic impact, i.e. prevent it from posing a threat to the existing theologico-ethical constellation. No wonder those who come closest to meeting these demands are neo-Kantians: Kant himself was focused on the problem of how, while fully taking Newtonian science into account, one can guarantee that ethical responsibility can be exempted from the reach of science – as he himself put it, he limited the scope of knowledge to create the space for faith and morality. And are today's state philosophers not facing the same task? Are their efforts not focused on how, through different versions of transcendental reflection, to restrict science to its preordained horizon of meaning and thus denounce as 'illegitimate' its consequences for the ethico-religious sphere? In this sense, Habermas is effectively the ultimate philosopher of (re)normalization, desperately working to prevent the collapse of our established ethico-political order:

> Could it be that Jurgen Habermas' corpus will be one day of the first in which simply nothing at all prodding can be found any more? Heidegger, Wittgenstein, Adorno, Sartre, Arendt, Derrida, Nancy, Badiou, even Gadamer, everywhere one stumbles upon dissonances. Normalization takes hold. The philosophy of the future – integration brought to completion.[4]

The reason for this Habermasian aversion to Sloterdijk is thus clear: Sloterdijk is the ultimate 'prodder', the one who is not afraid to 'think dangerously' and to question the presuppositions of human freedom and dignity, of our liberal welfare state, etc. One should not be afraid to call this orientation 'evil' – if one understands 'evil' in the elementary sense outlined by Heidegger: 'The evil and therefore most acute danger is thinking itself, insofar as it has to think against itself, yet can seldom do so.'[5] One should push Heidegger a step further here: it is not just that thinking is evil insofar as it fails to think against itself, against the accustomed way of thinking; thinking, insofar as its innermost potential is to think freely and 'against itself', is what, from the standpoint of conventional thinking, cannot but appear as 'evil'. It is crucial to persist in this ambiguity, as well as to resist the temptation to find an easy way out by defining some kind of 'proper measure' between the two extremes of normalization and the abyss of freedom.

Does this mean that we should simply choose our side in this opposition – 'corrupting the youth' or guaranteeing meaningful stability? The problem is that, today, simple opposition gets complicated: our global-capitalist reality, impregnated as it is by sciences, is itself 'prodding', challenging our innermost presuppositions in a much more shocking way than the wildest philosophical speculations, so that the task of a philosopher is no longer to undermine the hierarchical symbolic edifice that grounds social stability but – to return to Badiou – to make the young perceive the dangers of the growing nihilist order that presents itself as the domain of new freedoms. We live in an extraordinary era in which there is no tradition on which we can base our identity, no frame of meaningful universe which might enable us to lead a life beyond hedonist reproduction. Today's nihilism – the reign of cynical opportunism accompanied by permanent anxiety – legitimizes itself as the liberation from the old constraints: we are free to constantly re-invent our sexual identities, to change not only our job or our professional trajectory but even our innermost subjective features like our sexual orientation. However, the scope of these freedoms is strictly prescribed by the coordinates of the existing system, and also by the way consumerist freedom effectively functions: the possibility to choose and consume imperceptibly turns into a superego *obligation* to choose. The nihilist dimension of this space of freedoms can only function in a permanently accelerated way – the moment it slows down, we become aware of the meaninglessness of the entire movement. This New World Disorder, this gradually emerging world-less civilization, exemplarily affects the young, who oscillate between the intensity of fully burning out (sexual enjoyment, drugs, alcohol, even violence), and the endeavour to succeed (study, make a career, earn money . . . within the existing capitalist order). Permanent transgression thus becomes the norm – consider the deadlock of sexuality or art today: is there anything more dull, opportunistic or sterile than to succumb to the superego injunction to incessantly invent new artistic transgressions and provocations (the performance artist masturbating on stage or masochistically cutting himself, the sculptor displaying decaying animal corpses or human excrement), or to the parallel injunction to engage in more and more 'daring' forms of sexuality?

The only radical alternative to this madness appears to be the even worse madness of religious fundamentalism, a violent retreat into some artificially resuscitated tradition. The supreme irony is that a brutal return to an orthodox tradition (an invented one, of course) appears as the ultimate 'prodding' – are the young suicide bombers not the most radical form of corrupted youth? The great task of thinking today is to discern the precise contours of this deadlock and find the way out of it. A recent incident illustrates perfectly the paradoxical coincidence of opposites that underlies the retreat from fidelity to tradition into transgressive 'prodding'. In a hotel in Skopje, Macedonia, where I recently stayed, my companion enquired whether smoking was permitted in our room, and the answer she got from the receptionist was priceless: 'Of course not, it is prohibited by the law. But you have ashtrays in the room, so this is not a problem.' The contradiction between prohibition and permission was openly assumed and thereby cancelled, treated as non-existent: the message was, 'It's prohibited, and here is how you do it.' This incident perhaps provides the best metaphor for our ideological predicament today.

How did we reach this point? One of the greatest contributions of American culture to dialectical thinking is the series of rather vulgar doctor's jokes of the type, 'first-the-bad-news-then-the-good-news', like: 'The bad news is that you have terminal cancer and will die in a month. The good news is that we also discovered you have severe Alzheimer's, so you will already have forgotten the bad news when you get home.' Maybe we should adopt a similar approach to radical politics. After so much 'bad news' – seeing so many hopes brutally crushed in the space of radical action, spread between the two extremes of Maduro in Venezuela and Tsipras in Greece – it is easy to succumb to the temptation to claim that such action never really had a chance, that it was doomed from the very beginning, that the hope of a real and effective change for the better was a mere illusion. What we should do is not search for alternative 'good news' but discern the good news in the bad news, by way of changing our standpoint and seeing it in a new way. Take the prospect of automatization of production, which will – so people fear – radically diminish the need for workers and thus make unemployment explode. But why fear this prospect? Does it not open up the possibility of a new society in which we all have to work much less? In what kind of society do we live, where good news is automatically turned into bad news? Or, to take another example of bad/good news: is the basic lesson of the recent public disclosure of the so-called Paradise Papers not the simple fact that the ultra-rich live in their special zones where they are not bound by common laws?

New areas of emancipatory activity are emerging, such as those cities run by a mayor or city-council imposing progressive agendas that run counter to larger state or federal regulations. Examples abound here, from single cities (Barcelona, Newark, New York, even) to a network of cities – recently, many local authorities in the US decided to continue to honour commitments to fight ecological threats that were cancelled by the Trump administration. The important fact here is that local authorities proved to be more sensitive to global issues than higher state authorities. This is why we should not reduce this new phenomenon to the struggle of local communities against state regulations: local administrative authorities are concerned with issues that are simultaneously local and global, putting pressure on the state from two directions. For example, the mayor of

Barcelona insists on opening up the city to refugees, while she opposes the excessive invasion of tourists into the city.

Another emancipatory step is that women are coming out *en masse* about male sexual violence. The media coverage of this development should not distract us from what is really going on: nothing less than an epochal change, a great awakening, a new chapter in the history of equality. For thousands of years, relations between the sexes were regulated and arranged; all this is now being questioned and undermined. And now the protesters are not an LGBT+ minority but a majority – women. What is emerging is something we have been aware of all along but were just not able (willing, ready) to address openly: the hundreds of ways in which women are exploited sexually. Women are now drawing attention to the dark underside of our official claims of equality and mutual respect, and what we are discovering is, among other things, how hypocritical and one-sided our fashionable critique of women's oppression in Muslim countries is: we must confront the reality of our own forms of oppression and exploitation.

As in every revolutionary upheaval, there will be numerous 'injustices', ironies, and so on. (For example, I doubt that the American comedian Louis CK's acts, deplorable and lewd as they are, could be put on the same level as direct sexual violence.) But, again, none of this should distract us; rather, we should focus on the problems that lie ahead. Although some countries are already experiencing a new post-patriarchal sexual culture (look at Iceland, where two thirds of children are born out of a wedlock, and where women occupy more posts in public institutions than men), one of the most urgent tasks is to explore what we are gaining and losing in the upheaval of traditional courtship procedures. New rules will have to be established in order to avoid a sterile culture of fear and uncertainty – plus, of course, we must make sure that this awakening does not turn into just another case where political legitimization is based on the subject's victimhood status.

Is the basic characteristic of today's subjectivity not the weird combination of the free subject who experiences himself as being ultimately responsible for his fate, and the subject who grounds the authority of his speech on his status as a victim of circumstances beyond his control? Every contact with another human being is experienced as a potential threat – if the other smokes, or if he casts a covetous glance at me, he already hurts me. This logic of victimization is today universalized, reaching well beyond the standard cases of sexual or racist harassment – recall, for example, the growing financial industry of paying damages, from the tobacco companies' deal in the USA and the financial claims of the Holocaust victims and forced labourers in Nazi Germany, to the idea that the USA should pay African-Americans hundreds of billions of dollars for all they were deprived of due to slavery. This notion of the subject as an unresponsible victim is driven by an extreme narcissistic perspective in which every encounter with the Other appears as a potential threat to the subject's precarious imaginary balance; as such, it is not the opposite of, but rather the inherent supplement to, the liberal free subject. In today's predominant form of individuality, the self-centered assertion of the psychological subject paradoxically overlaps with the perception of oneself as a victim of circumstances.

To return to the ashtray: the danger is that, in a homologous way, in the ongoing awakening, the ideology of personal freedom could silently merge with the logic of victimhood (freedom being reduced to the freedom to bring out one's victimhood). A radical, emancipatory politicization of the awakening will then be superfluous and the women's fight will become one in a series of protests – against global capitalism, ecological threats, racism, for a different democracy, and so on.

So how will radical social transformation happen? Definitely not as a triumphant victory or even in the sort of catastrophe widely debated and predicted in the media, but 'as a thief in the night': 'For yourselves know perfectly that the day of the Lord so cometh as a thief in the night. For when they shall say, Peace and safety; then sudden destruction cometh upon them, as travail upon a woman with child; and they shall not escape' (Paul, 1 Thessalonians 5:2–3). Is this not already happening in our society, obsessed as it is with 'peace and security'? On a closer look, however, we see that the change is already happening in broad daylight: capitalism is openly disintegrating and changing into something else. We do not perceive this ongoing transformation because of our deep immersion in ideology.

The same holds for psychoanalytic treatment, where resolution also comes 'as a thief in broad daylight', as an unexpected by-product, never as the achievement of a posited goal. This is why psychoanalytic practice is something that is possible only because of its own impossibility – a statement which many would instantly proclaim a typical piece of postmodern jargon. However, did Freud himself not point in this direction when he wrote that the ideal conditions for psychoanalytic treatment would be those in which psychoanalysis is no longer needed? This is the reason why Freud listed the practice of psychoanalysis among the impossible professions. After psychoanalytic treatment begins, the patient resists it by (among other ways) deploying transferences, and the treatment progresses through the analysis of transference and other forms of resistance. There can be no direct, 'smooth' treatment: in a treatment, we immediately stumble upon obstacles by way of working through these obstacles.

And, back to politics: doesn't exactly the same hold for every revolution and every process of radical emancipation? Revolutions are only possible against the background of their own impossibility: the existing global-capitalist order can immediately counter all attempts to subvert it, and anti-capitalist struggle can only be effective if it deals with these countermeasures, if it turns into its weapon the very instruments of its defeat. There is no point in waiting for the right moment when a smooth change might be possible; this moment will never arrive, history will never provide us with such an opportunity. One has to take the risk and intervene, even if reaching the goal appears (and is, in some sense) impossible – only by doing this can one change the situation so that the impossible becomes possible, in a way that can never be predicted.

Although it may appear that we are hopelessly at the mercy of media manipulation,[6] miracles can happen, the fake universe of manipulation can all of a sudden crumble and undo itself. In the campaign that preceded the 2017 UK General Election, Jeremy Corbyn was the target of a well-planned character assassination by the conservative media, which portrayed him as undecided, incompetent, non-electable, and so on. So how did he emerge so well out of it? It is not enough to say that he successfully resisted the smears with his display of simple honesty, decency and concern for the worries of ordinary people. He did well precisely because of the attempted character assassination: without it he would probably remain a slightly boring and uncharismatic leader lacking a clear vision, merely a representative of the old Labour Party. It was in his reaction to the ruthless campaign against him that his ordinariness emerged as a positive asset, as something that attracted voters disgusted by the vulgar attacks on him, and this shift was unpredictable: it was impossible to determine in advance how the negative campaign would work. This undecidability (to use a once-fashionable word) is a feature of symbolic determination which cannot be accounted for in terms of simple linear determinism: it is not a question of insufficient data, of some arguments being stronger than others, but one of how the same arguments can work *for* or *against*. A character trait – Corbyn's accentuated ordinary decency – may be an argument for him (for the voters tired of the Conservative media blitz) or an argument against him (for those who think that a leader should be strong and charismatic). The added *je ne sais quoi* which decides how events will play out is what escapes the well-prepared propaganda.

Those who follow obscure spiritual-cosmological speculations will be familiar with a popular idea: when three planets (usually Earth, its moon and the Sun) find themselves along the same axis, some big cataclysmic event takes place; the whole order of the universe is momentarily thrown out of sync and has to restore its balance (as was supposed to happen in 2012). Did something like this not hold for the year 2017, which was a triple anniversary: in 2017 we celebrated not only the centenary of the October Revolution but also the one hundred and fiftieth anniversary of the first edition of Marx's *Capital* (1867), and the fiftieth anniversary of the so-called Shanghai Commune when, during the Cultural Revolution, the residents of Shanghai decided to follow literally Mao's call and directly took power, overthrowing the rule of the Communist Party (which is why Mao quickly decided to restore order by sending the army to squash the Commune). Do these three events not mark the three stages of the Communist movement: Marx's *Capital* outlined the theoretical foundations of the Communist revolution, the October Revolution was the first successful attempt to overthrow a bourgeois state and build a new social and economic order, while the Shanghai Commune stands for the most radical attempt to realize the most daring aspect of the Communist vision, the abolition of state power and the imposition of direct people's power, organized as a network of local communes.

The lesson here is that, when we are considering the centenary of the October Revolution – the first case of a 'liberated territory' outside capitalism, of taking power and breaking the chain of capitalist states – we should always see it as the middle (mediating) stage between two extremes, the antinomic structure of the capitalist society (analysed in *Capital*), out of which the Communist movement grew, and the no less antinomic *péripéties* of Communist state power, which culminated in the *cul de sac* of the Chinese Cultural Revolution. After taking over, the new power confronts the immense task of organizing the new society. Remember the exchange between Lenin and Trotsky on the eve of the October Revolution: Lenin said, 'What will happen to us if we fail?' Trotsky replied: 'And what will happen if we succeed?'

Today, we are stuck with this question. The present book deals with it in three tragic acts plus a fourth one, a sort of comic supplement. The book's premise is that today, more than ever, we should stick to the basic Marxist insight: Communism is not an ideal, a normative order, a kind of ethico-political 'axiom', but something that arises as a reaction to the ongoing historical process and its deadlocks. Back in 1985, Félix Guattari and Toni Negri published a short book in French called *Les nouveaux espaces de liberté*, whose title was changed for the English translation into *Communists Like Us* (Los Angeles: Semiotexte 1990)[7] – in an unintended way, this title points to the forthcoming upper-middle-classization of the Communist idea, which made a modest return as a slogan for some well-to-do academics with no connection to the actual poor and exploited. The new Communists are 'like us', ordinary academic cultural Leftists; there is no radical subjective transformation involved. 'Communism' becomes an island to which one 'subtracts' oneself – a nice case of what one can call 'principled opportunism', i.e. sticking faithfully to abstract 'radical' notions as a way to remain 'pure', avoiding 'compromises' because one also avoids any engagement in actual politics.

So when we talk about the continuing relevance (or irrelevance, for that matter) of the idea of Communism, we should not be thinking of a regulative idea in the Kantian sense but in the strict Hegelian sense – for Hegel, 'idea' is a concept which is not a mere Ought (*Sollen*) but also contains the power of its actualization. The question of the actuality of the idea of Communism is therefore that of discerning in our actuality tendencies which point towards it, otherwise it's an idea not worth losing time with.

1
The State of Things

The Topsy-Turvy World of Global Capitalism

To really change things, one should accept that nothing can really be changed within the existing system. Jean-Luc Godard voiced the motto, 'Ne change rien pour que tout soit différent' ('Change nothing so that everything will be different'), a reversal of 'Some things must change so that everything remains the same'. In our late-capitalist consumerist dynamic we are bombarded by new products all the time, but this constant change is becoming increasingly monotonous. When only constant self-revolutionizing can maintain the system, those who refuse to change anything are effectively the agents of true change: a change to the very principle of change.

Or, to put it in a different way, true change is not just the overthrowing of the old order but, above all, the establishment of a new order. Louis Althusser once improvised a typology of revolutionary leaders worthy of Kierkegaard's classification of humans into officers, housemaids and chimney sweepers: those who quote proverbs, those who do not quote proverbs, and those who invent new proverbs. The first are scoundrels (Althusser thought of Stalin), and the second are great revolutionaries who are doomed to fail (Robespierre); only the third understand the true nature of a revolution and succeed (Lenin, Mao). This triad registers three different ways in which to relate to the big Other (the symbolic substance, the domain of unwritten customs and wisdoms best expressed in the stupidity of proverbs). Scoundrels simply reinscribe the revolution into the ideological tradition of their nation (for Stalin, the Soviet Union was the last stage of the progressive development of Russia). Radical revolutionaries like Robespierre fail because they merely enact a break with the past without succeeding in their effort to enforce a new set of customs (recall the utmost failure of Robespierre's idea to replace religion with the new cult of a Supreme Being). Leaders like Lenin and Mao succeeded (for some time, at least) because they invented new proverbs, which means that they imposed new customs that regulated daily lives. One of the best Goldwynisms recounts how, after being told that critics had complained that there were too many old clichés in his films, Sam Goldwyn wrote a memo to his scenario department: 'We need more new clichés!' He was right, and this is a revolution's most difficult task – to create 'new clichés' for ordinary daily life.

One should take a step further here. The task of the Left is not just to propose a new order, but also to change the prospect of what appears possible. The paradox of our predicament is therefore that, while resistance to global capitalism seemingly fails again and again to halt its advance, it fails to recognize the many trends which clearly signal capitalism's progressive disintegration. It is as if the two tendencies (resistance and self-disintegration) move at different levels and cannot meet, so that we get futile protests at the same time as immanent decay and there is no way of bringing the two together in a coordinated attempt to emancipate the world from capitalism. How did it come to this? While most of the Left desperately try to protect workers' rights against the onslaught of global capitalism, it is almost exclusively the most 'progressive' capitalists themselves (from Elon Musk to Mark Zuckerberg) who talk about post-capitalism – as if the very concept of the passage from capitalism as we know it to a new post-capitalist order is being appropriated by capitalism itself.

In an interview for *The Atlantic* in November 2017, Bill Gates said that capitalism isn't working, and that socialism is our only hope in order to save the planet. His reasoning is based on a simple ecological calculation: the use of fossil fuels has to be radically reduced if we are to avoid a global catastrophe, and the private sector is too selfish to produce clean and economical alternatives, so humanity has to act outside market forces. Gates himself announced his intention to spend $2 billion of his own money on green energy, although there's no fortune to be made in it, and he called on fellow billionaires to help make the US fossil-free by 2050 with similar philanthropy.[1] From an orthodox Leftist position, it is easy to make fun of the naivety of Gates's proposal. Such reproaches might be right, but they raise the following question: where is the Left's realistic proposal as to what we should do? Words matter in public debates: even if what Gates is talking about is not 'true socialism', he does talk about the fateful limitation of capitalism – and, again, do today's self-proclaimed socialists have a serious vision of what socialism should be now?

The standard radical Leftist reproach to the Left's record in power is that, instead of effectively socializing production and deploying actual democracy, it remained within the constraints of conventional Leftist policies (nationalizing the means of production or tolerating capitalism in a social-democratic way, imposing an authoritarian dictatorship or playing the game of parliamentary democracy). Maybe the time has come to ask the brutal question: OK, but what should or could they have done? How would an authentic model of socialist democracy have looked in practice? Is this Holy Grail – a revolutionary power that avoids all the traps (Stalinism, Social Democracy) and develops an authentic people's democracy in terms of society and the economy – not a purely imaginary entity, one which by definition cannot be filled with actual content?

Hugo Chávez, President of Venezuela from 1999 to 2013, was not simply a populist throwing the oil money around. Largely ignored by the international media are the complex and often inconsistent efforts to overcome a capitalist economy by experimenting with new ways of organizing production, ones which endeavour to move beyond the alternatives of private or state-owned property: farmers' and workers' cooperatives, workers' participation, control and organization of production, different hybrid forms between private property and social control and organization, and so on. Factories not used by their owners might be given to the workers to run, say. There are many hits and misses on this path – for example, after several attempts, the plan to hand over nationalized factories to workers, distributing stocks among them, was abandoned. Although these are genuine efforts to integrate grass-roots initiatives with state proposals, one must also note the many economic failures and inefficiencies and the widespread corruption that took place. The usual story is that, after half a year of enthusiastic work, things go downhill. In the first years of Chávismo, we were clearly witnessing a broad popular mobilization. However, the big question remains: how does or should this reliance on popular self-organization affect running a government? Can we even imagine today an authentic Communist power? What we get is disaster (Venezuela), capitulation (Greece), or a full return to capitalism (China, Vietnam).

Official attempts at Marxist social theory in China try to paint a picture of today's world which, to put it simply, basically remains the same as that of the Cold War years: the worldwide struggle between capitalism and socialism goes on unabated, the fiasco of 1990 was just a temporary setback, so that today the big opponents are no longer the USA and the USSR but the USA and China, which remains a socialist country. The explosion of capitalism in China is viewed as a monumental case of what in the early Soviet Union they called the New Economic Policy, so that what we have in China is a new 'socialism with Chinese characteristics', but still socialism: the Communist Party remains in power and tightly controls and directs market forces. From this standpoint, the economic success of China in the last decades is interpreted as proof not of the productive potential of capitalism but of the superiority of socialism over capitalism. To sustain this view, which also counts Vietnam, Venezuela, Cuba and even Russia as socialist countries, one has to give new socialism a strong socially conservative twist. This is not the only reason why the rehabilitation of socialism is blatantly non-Marxist, totally ignoring the basic Marxist point that capitalism is defined by capitalist relations of production, not by the type of state power.[2]

All those who have any illusions in Putin should note the fact that he elevated to the status of his official philosopher one Ivan Ilyin, a Russian political theologist who, after being expelled from the Soviet Union in the early 1920s on the famous 'philosophers' steamboat', advocated, against both Bolshevism and Western liberalism, his own version of Russian Fascism: the state as an organic community led by a paternal monarch. One must nonetheless concede a partial truth in this Chinese position: even in the wildest capitalism, it matters who controls the state apparatuses. Classical Marxism and the ideology of neo-Liberalism both tend to reduce the state to a secondary mechanism that obeys the needs of the reproduction of capital; they both thereby underestimate the active role played by state apparatuses in economic processes. Today one should not fetishize capitalism as the Big Bad Wolf that is controlling states: state apparatuses are active in the very heart of economic processes, doing much more than just guaranteeing legal and other (educational, ecological) conditions of the reproduction of capital. In many different forms, the state is active as a direct economic agent (it helps failing banks, it supports selected industries, it orders defence and other equipment) – in the US today, around 50 per cent of production is mediated by the state (while a century ago, this percentage was between 5 and 10). Marxists should have learned this lesson from state Socialism, where the state was a direct economic agent and regulator, so that, whatever it was, it was a state without a capitalist class – certain Marxist analysts use the suspicious term 'state capitalism' to account for it. But if we can get a capitalist state without capitalists as a class, to what extent can we imagine a non-capitalist state with capitalists playing a strong role in the economy? While the Chinese model is certainly inadequate – it combines exploding social inequalities with a strong authoritarian state – one should nonetheless not exclude *a priori* the possibility of a strong non-capitalist state that resorts to elements of capitalism in some of the domains of social life. It is possible to tolerate limited elements of capitalism without allowing the logic of capital to become the overdetermining principle of the social totality.

As Julia Buxton puts it, the Bolivarian revolution 'has transformed social relations in Venezuela and had a huge impact on the continent as a whole. But the tragedy is that it was never properly institutionalized and thus proved to be unsustainable.'[3] OK, but how to institutionalize it in an authentic way? It is all too easy to say that authentic emancipatory politics should remain at a distance from the state; the fundamental problem is what to do with that state. Can we even imagine a society outside the state? We must deal with these problems here and now: there is no time to wait for some future solution and, in the meantime, keep a safe distance from the state. In other words, why was there no Venezuelan Left to provide an authentic radical alternative to Chávez and Maduro? Why was the initiative in opposing Chávez ceded to the extreme Right, which triumphantly hegemonized the oppositional struggle, claiming to be the voice of the ordinary people who suffered the consequences of Chávez's mismanagement of the economy?

In early March 2018, a small news item passed almost unnoticed among the 'big' events: in South Africa, the ruling party (the African National Congress) decided to dispossess the white farmers of their land without compensation. This decision – if it is realized – will again confront the Left with a big dilemma. Obviously, something has to be done, since the white minority still owns most of the arable land as the result of apartheid. However, how would such a measure be realized without causing another Zimbabwe-like economic catastrophe, which would play into the hands of the liberal opinion that Blacks cannot really run an economy, and also discredit radical Leftist measures in general?

In short, what if the search for an authentic Third Way – beyond Social Democracy, which doesn't go far enough, and 'totalitarianism', which goes too far – is a waste of time? The strategy of the radical Left is to try to demonstrate, with great theoretical sophistication, how 'totalitarian' radicalization masks its opposite: Stalinism was effectively a form of state capitalism, and so on. In the case of Venezuela, radical Leftists blame the fiasco of Chávismo on the fact that it compromised with capitalism, not only by drowning in corruption but by making deals with international corporations to exploit Venezuela's natural resources. Again, while this is in principle true, what should the government have done? In Bolivia the Morales–Linera government avoided these pitfalls, but did they do anything more than remain within the confines of a more modest, 'democratic' form of politics?

Perhaps, in order to break out of this deadlock, the first step should be to drop our obsession with progress and instead focus on those who are left behind – by gods and by the market. One unexpected topic has emerged in popular fiction in the last decades, from the lowest fundamentalist trash (Tim LaHaye *et consortes*) to TV series (*Leftovers*) – the issue of those 'left behind'. Armageddon is approaching, and God has taken to himself the privileged ones in order to save them from the forthcoming horrors. But what if we propose a vulgar economic reading of the popular appeal of this idea? As is often the case, it seems that God himself listened to the voice of Capital, so the question of those left behind is related to our economic predicament in global capitalism. Is it, surely, not only those who were unable to join the flow of refugees and had to remain stuck in their country in disarray who are the 'left behind'?

Most refugees don't want to live in Europe, they want a decent life back home. Instead of working to achieve that, Western powers treat the problem as a 'humanitarian crisis' whose two extremes are hospitality and the fear of losing our way of life. They thereby create a pseudo-'cultural' antagonism between refugees and the local lower-class population, engaging them in a conflict which transforms a politico-economic struggle into one of the 'clash of civilizations'.

One should avoid any simplistic romanticization of refugees. Some European Leftists claim that refugees are a nomadic proletariat which could act as the core of a new revolutionary movement in Europe – a claim that is deeply problematic. The proletariat is, for Marx, composed of exploited workers disciplined through work and creating wealth, and, while today's precariat can count as a new form of proletariat, the paradox of refugees is that they are mostly seeking to become the proletariat. They are 'nothing', with no place within the social hierarchy of the country where they take refuge; but from here it is a big step to becoming a proletariat in the strict Marxian sense. So instead of celebrating refugees as nomadic proletarians, would it not be more appropriate to claim that they are the more dynamic and ambitious part of their country's population, those with a will to achieve, and that the true proletarians are rather those who remained and were left behind as strangers in their own country (with all the religious connotations: 'leftovers', those not taken to God in rapture).

The tendency of global capitalism is to make 80 per cent of us 'left behind'. A century ago, Vilfredo Pareto was the first to identify the so-called 80/20 rule: 80 per cent of the land is owned by 20 per cent of the people, 80 per cent of profits are produced by 20 per cent of the employees, 80 per cent of decisions are made during 20 per cent of meeting time, 80 per cent of the links on the Web point to less than 20 per cent of Web pages, 80 per cent of peas are produced by

20 per cent of the pea pods. As some social analysts and economists have suggested, today's explosion of economic productivity confronts us with the ultimate case of this rule: the coming global economy tends towards a state in which only 20 per cent of the workforce can do all the necessary jobs, so that 80 per cent of the people are basically irrelevant and of no use, potentially unemployed. When this logic reaches its extreme, would it not be reasonable to reduce it to its self-negation: is not a system which makes 80 per cent of the people irrelevant and of no use *itself irrelevant and of no use*? The problem is thus not primarily that a new global proletariat is emerging, but something much more radical: billions of people are simply not needed, the sweatshops cannot absorb them. This aspect is neglected by Leftist politics, which is reduced to fighting to conserve the fast-disappearing remains of the welfare state; but with the ongoing devastating economic politics this is a lost fight. Lost not simply because of the financial elite which profits from its loss, but because this same financial elite can rely on the growing army of those who *never even had* access to any of these 'benefits' and instead denounce them as privileges (young, precarious workers).[4]

The fight to maintain old welfare state benefits is thus ultimately that of the established working class against the new over-exploited marginals (precarious workers, new slaves, etc.), who never even enjoyed these benefits. The Italian Marxist Toni Negri once gave an interview, strolling along a suburban street in Venezia-Mestre; the journalist's camera caught him passing a line of workers picketing in front of a textile factory which was due to be closed down. He pointed at the workers and dismissively remarked: 'Look at them! They don't know they are already dead!' For Negri, these workers stood for all that is wrong with the traditional trade-unionist socialism focused on corporate job security, a socialism mercilessly rendered obsolete by the dynamics of 'postmodern' capitalism with its hegemony over intellectual labour. Instead of reacting to this new 'spirit of capitalism' in the way traditional Social Democracy does, seeing it as a threat, Negri claims we should fully embrace it, discerning in it – in the dynamics of intellectual labour and its non-hierarchic and non-centralized social interaction – the seeds of communism. If we follow this logic to its conclusion, we cannot but agree with the cynical quip that, today, the main task of trade unions should be to re-educate workers so that they will be able to adapt to the new digitalized economy. The problem with Negri's vision resides in his use of the Spinozan notion of multitude: rather than accepting his own enthusiastic examples and those of Michael Hardt, we should recall the last scene (in the original 1872 version) of Mussorgsky's *Boris Godunov*, where, in a forest glade near Kromï, a crowd celebrates the fall of the Tsar. Here is the description (shamelessly taken from Wikipedia):

> Tempestuous music accompanies the entry of a crowd of vagabonds who have captured the boyar Khrushchov. The crowd taunts him, then bows in mock homage ('Not a falcon flying in the heavens'). The yuródiviy enters, pursued by urchins. He sings a nonsensical song ('The moon is flying, the kitten is crying'). The urchins greet him and rap on his metal hat. The yuródiviy has a kopek, which the urchins promptly steal. He whines pathetically. Varlaam and Misail are heard in the distance singing of the crimes of Boris and his henchmen ('The sun and moon have gone dark'). They enter. The crowd gets worked up to a frenzy ('Our bold daring has broken free, gone on a rampage'), denouncing Boris. Two Jesuits are heard in the distance chanting in Latin ('Domine, Domine, salvum fac'), praying that God will save Dmitriy. They enter. At the instigation of Varlaam and Misail, the vagabonds prepare to hang the Jesuits, who appeal to the Holy Virgin for aid. Processional music heralds the arrival of Dmitriy and his forces. Varlaam and Misail glorify him ('Glory to thee, Tsarevich!') along with the crowd. The Pretender calls those persecuted by Godunov to his side. He frees Khrushchov, and calls on all to march on Moscow. All exeunt except the yuródiviy, who sings a plaintive song ('Flow, flow, bitter tears!') of the arrival of the enemy and of woe to Russia.

In this chaotic mixture of voices (Orthodox believers, Catholic emissaries, the pretender Dmitriy and his propagandists, a frightened boyar, sadistically playful children), compassion is combined with opportunism, innocence with corruption, passion for freedom with deft manipulation. We are here as far as imaginable from any assertion of emancipatory popular will – what lurks in the background is an impenetrable darkness. But what about the opposite vision? Insofar as the dynamics of new capitalism are rendering an increasingly large percentage of workers superfluous, what about the project of reuniting all the 'living dead' of global capitalism, all those left behind by neo-capitalist 'progress', all those rendered superfluous, obsolete, all those unable to adapt to new conditions? The wager is, of course, that a direct short-circuit can be created between these leftovers of history and history's most progressive aspects.[5]

Liberals who acknowledge the problems of those excluded from the socio-political process see their goal as the inclusion of those whose voices are not heard: all points of view should be listened to, all interests taken into account, the human rights of everyone guaranteed, all ways of life, cultures and practices respected. The obsession of this form of democracy is the protection of all kinds of minorities: cultural, religious, sexual, etc. The formula of democracy here is patient negotiation and compromise. What gets lost is the proletarian position, that of universality embodied in the excluded. This is why, upon a closer look, it becomes clear that liberal inclusion was not what Chávez was trying to achieve: he was not including the left-behind in the pre-existing liberal-democratic framework; he was, on the contrary, taking them, the 'excluded' dwellers of favelas, as his basis, and reorganizing politics to fit around them. Pedantic and abstract as it may appear, this difference – the one between 'bourgeois democracy' and 'dictatorship of the proletariat' – is crucial.

The true choice is, therefore: should we continue to play the humanitarian game of taking care of those left behind, or should we tackle the much more difficult task of changing the global system that generates them? Without such a change, our situation will be increasingly irrational. In order to orient ourselves in this conundrum, we should be aware of the fateful limitation of the politics of interests. Parties like Die Linke in Germany represent the interests of their working-class constituency – better healthcare and retirement conditions, higher wages, and so on; this automatically puts them within the confines of the existing system, and their goal is therefore not authentic emancipation. Interests are not merely to be followed; they have to be redefined to accommodate ideas that cannot be reduced to interests.[6] This is why we witness, again and again, the paradox of Rightist populists, when they attain power, sometimes imposing measures which are in workers' interests – as is the case in Poland, where:

> PiS [Law and Justice, the ruling Rightist-populist party] has transformed itself from an ideological nullity into a party that has managed to introduce shocking changes with record speed and efficiency . . . it has enacted the largest social transfers in Poland's contemporary history. Parents receive a 500 złoty ($120) monthly benefit for every child after their first, or for all children in poorer families (the average net monthly income is about 2,900 złoty, though more than two thirds of Poles earn less). As a result, the poverty rate has declined by 20–40 per cent, and by 70–90 per cent among children. The list goes on: in 2016, the government introduced free medication for people over the age of 75. The minimum wage now exceeds what trade unions had sought. The retirement age has been reduced from 67 for both men and women to 60 for women and 65 for men. The government also plans tax relief for low-income taxpayers.[7]

PiS stands for what Marine Le Pen has also promised to do in France: a combination of anti-austerity measures – social transfers no Leftist party dares to consider – plus the promise of order and security that asserts national identity and pledges to deal with the immigrant threat. Who can beat this combination, which directly addresses the two big worries of ordinary people? We can discern on the horizon a weirdly perverted situation in which the official 'Left' is enforcing austerity politics (at the same time advocating multicultural rights), while the populist Right is implementing anti-austerity measures to help the poor (while pursuing a xenophobic nationalist agenda) – the latest example of what Hegel described as '*die verkehrte Welt*', the topsy-turvy world.

Virtual Capitalism and the End of Nature

Although Marx provided an unsurpassable analysis of capitalist reproduction, his mistake was not just that he counted on capitalism's final breakdown, and therefore couldn't grasp how capitalism came out of each crisis strengthened; he made a much more fundamental error. Marxism is described in precise terms by Wolfgang Streeck: Marxism was right about the 'final crisis' of capitalism, and we are clearly entering it today; but this crisis is precisely that, a prolonged process of decay and disintegration with no easy Hegelian *Aufhebung* in sight, no agent to give this decay a positive twist and make it the means of some higher level of social organization:

> It is a Marxist – or better: modernist – prejudice that capitalism as a historical epoch will end only when a new, better society is in sight, and a revolutionary subject ready to implement it for the advancement of mankind. This presupposes a degree of political control over our common fate of which we cannot even dream after the destruction of collective agency, and indeed the hope for it, in the neoliberal-globalist revolution.[8]

Streeck enumerates different signs of this decay: lower profit rates, increased corruption and violence, financialization (profit from financial dealings parasitic upon value production). The paradox of the financial politics of the US and EU is that gigantic inputs of money fail to generate production since they mostly disappear in the operations of fictitious capital. This is why one should reject the standard liberal Hayekian interpretation of exploding debt (the costs of supplying a welfare state): data clearly shows that the bulk of it goes to feed financial capital and its profits.

There is another unexpected consequence of this decay. Rebecca Carson[9] recently demonstrated how the financialization of capital (where most profit is generated in M-M, the exchange of money for money, without routing through the valorization [*Verwertung*] of the labour force, which produces surplus value) paradoxically leads to the return of direct personal relations of domination – unexpectedly, since (as Marx emphasized), M-M is capital at its most impersonal and abstract. It is crucial to grasp here the link between three elements: fictitious capital, personal domination and the social reproduction of labour power. Financial speculation takes place before valorization; it mostly consists of credit operations and speculative investments where no money is yet spent on investment in production. Credit means debt, and therefore the subjects of this operation (not just individuals, but banks and institutions that manage money) are not involved in the process merely as subjects of the value-form, but are also creditors and debtors and as such are also part of another form of power relation that is not based on the abstract domination of commodification:

> Hence, the particular power relation involved in credit operations has a personal dimension of dependency (credit-debt) that is differentiated from abstract domination. This personal power relation, however, comes into being by the very process of exchange that is described abstractly by Marx as completely impersonal and formal since the social relations of credit operations are built on the social relations of the value-form. Hence the phenomenon of personal forms of dependency coming to the fore by way of the suspension of valorization with fictitious capital does not mean that abstract forms of domination are not also present.[10]

The power dynamic implied by fictitious capital is not a straightforward dichotomy between agents: while personal domination by definition occurs at the level of direct interaction, debtors are mainly not individuals but banks and hedge funds speculating on future production. And, in effect, are the operations of fictitious capital not made more and more without any direct intervention at all, i.e. simply by computers acting on their programs? However, these operations somehow have to be retranslated into personal relations, and here abstraction appears as personal domination. If capital is financialized and increasingly fictitious, so that relations between people are less and less mediated through commodification, then what happens to these relations? There is only one way out: relations of direct domination have to return in some way.

Those who are not subjected to direct commodification but who play a crucial role in the reproduction of labour power are also affected by the growing dependence on future valorization that is supposed to be opened up by the circulation of fictitious capital: fictitious capital is upheld by the expectation that valorization will occur in the future, therefore the reproduction of labour power is put under pressure so that those not labouring in the present will be ready to labour in the future. This is why the topic of education (in its productive-technocratic version: getting ready for the competitive job market) is so important today, and is also intertwined with debt: a student gets into debt to pay for his/her education, and this debt is expected to be repaid through self-commodification, i.e. when the indebted student gets a job. Education also emerges as one of the main topics in dealing with refugees – how to make them into a useful workforce.

Since, in our society, free choice is elevated into a supreme value, social control and domination can no longer appear to infringe on the subject's freedom – it has to appear as (and be sustained by) the very experience of individuals as being free. There are a multitude of ways in which this un-freedom appears in the guise of its opposite: when we are deprived of universal healthcare, we are told that we are given a new freedom of choice (to choose our healthcare provider); when we can no longer rely on long-term employment and are compelled to search for a new, precarious position every couple of years, we are told that we are given the opportunity to reinvent ourselves and discover novel, unexpected creative potentials that lurk in our personality; when we have to pay for the education of our children, we are told that we become 'entrepreneurs of the self', acting like a capitalist who chooses freely how he will invest the resources he possesses (or has borrowed) in education, health, travel . . . Constantly bombarded by so-called 'free choices', forced to make decisions for which we are mostly not even properly qualified (or about which we possess inadequate information), increasingly we experience our freedom as what it effectively is: a burden that deprives us of the true choice of change. Bourgeois society generally obliterates castes and other hierarchies, equalizing all individuals as market subjects divided only by class difference; but today's late capitalism, with its 'spontaneous' ideology, endeavours to obliterate the class division itself by way of proclaiming us all 'self-entrepreneurs', the differences among us being merely quantitative (a big capitalist borrows hundreds of millions for his investment, a poor worker borrows a couple of thousand for his supplementary education).

The much-celebrated 'Collaborative Commons' also play a role here. Marx always emphasized that the exchange between worker and capitalist is 'just' in the sense that workers (as a rule) get paid the full value of their labour power as a commodity: there is no direct 'exploitation' here, it is not that workers 'are not paid the full value of the commodity they are selling to the capitalists'. So while, in a market economy, I remain *de facto* dependent, this dependency is nonetheless 'civilized', enacted in the form of a 'free' market exchange between me and other persons instead of direct servitude or even physical coercion. It is easy to ridicule Ayn Rand, but there is a grain of truth in the famous 'hymn to money' from her *Atlas Shrugged*: 'Until and unless you discover that money is the root of all good, you ask for your own destruction. When money ceases to become the means by which men deal with one another, then men become the tools of other men. Blood, whips and guns or dollars. Take your choice – there is no other.'[11] Did Marx not say something similar in his well-known formula of how, in the universe of commodities, 'relations between people assume the guise of relations among things'? In the market economy, relations between people can appear as ones of mutually recognized freedom and equality; domination is no longer directly enacted and visible as such. Socialism as it really existed in the twentieth century proved that overcoming market-alienation abolishes 'alienated' freedom, and with it freedom *tout court*, bringing us back to 'non-alienated' relations of direct domination. To what extent are Collaborative Commons exposed to the same danger? Can they survive without a regulating agency which controls the very medium of collaboration and thereby exerts direct domination?

This, of course, in no way implies that, in this new form of domination, money plays no role, that we are dealing with direct domination: money continues to play a crucial role, but insofar as its distribution is no longer grounded in the process of valorization (workers being paid for their labour, etc.), it begins to function as a direct means of domination. In other words, money is used as a means of political power, as a way to exert this power and control its subjects. Furthermore, although some theorists claim that we thereby move beyond relations of commodity exchange and exploitation-through-valorization, one should insist that valorization via the circulation of capital remains the ultimate goal of the entire process of economic reproduction.

The expected outcome is that other divisions and hierarchies emerge: experts and non-experts, full citizens and the excluded, religious, sexual and other minorities. All groups not yet included in the process of valorization, up to refugees and citizens of 'rogue countries', are thus progressively subsumed in forms of personal domination, from the organization of refugee camps to judicial control of those considered potential lawbreakers – a form of domination which often has a human face (like social services intended to ease the refugees' smooth 'integration' into our society).

Why this resurgence of direct, non-democratic authority? Above and beyond cultural differences, it is inherently necessary in the logic of today's capitalism. The central problem we are facing today is: how does the late-capitalist predominance (or even hegemonic role) of 'intellectual labour' affect Marx's basic scheme of the separation of labour from its objective conditions, as well as of revolution as the subjective re-appropriation of the objective conditions? In spheres like the www communication network, production, exchange and consumption are inextricably intertwined, potentially even identical: my product is immediately communicated to and consumed by another. Marx's classic notion of 'commodity fetishism', in which 'relations between people' assume the form of 'relations between things', thus has to be radically rethought: in 'immaterial labour', 'relations between people' are 'not so much hidden beneath the veneer of objectivity, but are themselves the very material of our everyday exploitation',[12] so we cannot any longer talk about 'reification' in the classic Lukácsian sense. Far from being invisible, social relationality, in its very fluidity, is the object of marketing and exchange: in 'cultural capitalism', one no longer sells (and buys) objects that 'bring' cultural or emotional experience, one directly sells (and buys) such experiences. And since social relationship is directly marketed, this means that personal relations of domination are, too – I pay others to act as my servants . . . No wonder that, to obfuscate this breach of freedom and restore a false balance, many top managers pay prostitutes to play masochistic games of self-humiliation with them.

The power of the market economy to reflexively appropriate resistance to itself seems inexhaustible. Every owner of smartphones has experienced the superego pressure to learn and use all of its possible applications, from Facebook and Twitter to recording and taking snapshots. After buying the expensive machine that promises unheard-of wonders, one cannot avoid feeling guilty for ignoring all this wealth and just using it as an ordinary phone. The possibility of activating new applications imperceptibly slips into the obligation to do so. And now, what had to happen has happened: this uneasiness of ours has been detected and we are offered a way out of it. The Light Phone 2

> brings a few essential tools, like messaging, an alarm clock, or a ride home, so it's even easier to ditch your smartphone more often, or for good. It's a phone that actually respects you.[13]

A weird negativity is at work here – you are buying the phone because of what you don't want it to do (you don't want it to tempt you into going on Facebook, Twitter etc.). You thus get caught in a circular paradox: first you pay for all the additional functions provided by smartphones, then you pay even more to acquire some freedom and get rid of these additional functions. 'Respect' is a strange word to be used here – it implies that smartphones in some sense actually don't respect you. It is crucial to note that they are not simply eliminated. We are not expected to throw our smartphone away: the 'dumb' Light Phone just enables us to gain some breathing space, to escape the hold of the smartphone temporarily, to leave it behind in order to spend some quality time . . . in short, the Light Phone only works properly if the threat of the smartphone continues to lurk in the background – if we just drop the smartphone and use only the Light Phone, we simply regress to a lower technological stage and thus return to stupidity.

All these complications compel us to rethink the so-called 'labour theory of value', which should in no way be read as claiming that we should reject exchange, or its role in the constitution of value, as a mere appearance which obscures the key fact that labour is the origin of value. If we consider money merely as a secondary resource, a practical means that facilitates exchange, then the door is open to the illusion, succumbed to by Leftist followers of Ricardo, that it would be possible to replace money with simple notes designating the amount of work done by their bearer and giving him or her the right to the corresponding part of the social product – as if, by means of this direct 'work money', one could avoid all 'fetishism' and ensure that each worker is paid his or her 'full value'. The point of Marx's analysis is that this project ignores the formal determinations of money that make fetishism a necessary effect. In other words, when Marx defines exchange value as the mode of appearance of value, we should mobilize the entire Hegelian weight of the opposition between essence and appearance: essence exists only insofar as it appears – it does not pre-exist its appearance. In the same way, the value of a commodity is not its intrinsic substantial property which exists independently of its appearance in exchange.

This is also why we should abandon the attempt to expand the definition of value so that all kinds of labour are recognized as a source of value – recall the great feminist demand of the 1970s to recognize housework (from cooking and maintaining the household to caring for children) as being productive of value, or the demands of some contemporary eco-capitalists to integrate the 'free gifts of nature' into value production (by way of trying to determine the costs of water, air, forests and all other commons). All these proposals are 'nothing more than a sophisticated green-washing and commodification of a space from which a fierce attack upon the hegemony of the capitalist mode of production and its (and our) alienated relation to nature can be mounted':[14] in their attempt to be 'just' and to eliminate or at least constrain exploitation, such actions merely enforce an even stronger, all-encompassing commodification. Although they try to be 'just' at the level of content (what counts as value), they fail to problematize the very *form* of commodification: value should be treated in dialectical tension with non-value, i.e. to assert and expand spheres not caught in the production of market value, like housework or 'free' cultural and scientific work, in their crucial role. Value production can only thrive if it incorporates its inherent negation, the creative work that generates no market value; it is by definition parasitic on it. So instead of commodifying exceptions and including them in the process of valorization, one should leave them outside and destroy the frame that makes their status inferior. The problem with fictitious capital is not that it is outside valorization but that it remains parasitic on the fiction of a valorization to come.

A further challenge to the market economy comes from the spreading virtualization of money, which compels us to thoroughly reformulate the standard Marxist concepts of 'reification' and 'commodity fetishism', insofar as this topic still relies on the notion of a fetish as a solid object whose stable presence obfuscates its social mediation. Paradoxically, fetishism reaches its acme precisely when the fetish itself is 'dematerialized', turned into a fluid, 'immaterial' virtual entity; money fetishism will culminate in its passage into electronic form, when the last traces of its materiality disappear. Electronic money is the third form, after 'real' money, which directly embodies its value (gold, silver), and paper money, which, although a 'mere sign' with no intrinsic value, still clings to its material existence. And it is only at this stage, when money becomes a purely virtual point of reference, that it finally assumes the form of an indestructible spectral presence: I owe you \$1,000, and no matter how many material notes I burn, I still owe you \$1,000, the debt is inscribed somewhere in the virtual digital space. It is only with this thorough 'dematerialization' that Marx's famous old thesis from *The Communist Manifesto* – according to which, in capitalism, 'all that is solid melts into air' – acquires a much more literal meaning than the one he had in mind, when not only our material social reality is dominated by the speculative movement of capital, but this reality itself is progressively 'spectralized' (the 'Protean Self' instead of the old self-identical subject, the elusive fluidity of its experiences instead of the stability of the owned objects) – in short, when the usual relationship between material objects and immaterial ideas is turned upside down (objects are progressively dissolved in fluid experiences, while the only stable things are virtual symbolic obligations). It is only at this point that what Derrida called the spectral aspect of capitalism is fully actualized.

However, as we have already seen, such spectralization of the fetish contains the seeds of its opposite, of its self-negation: the unexpected return of direct relations of personal domination. Capitalism legitimizes itself as the economic system that implies and furthers personal freedoms (as a condition of market exchange); but its own dynamics have brought about a renaissance of slavery. Although slavery became almost extinct at the end of the Middle Ages, it exploded again in the European colonies from early modern times until the American Civil War. Today, a new era of slavery is arising along with the new epoch of global capitalism. Although it no longer has a legal status, slavery acquires a multitude of new forms: millions of immigrant workers in the Saudi peninsula, who are deprived of elementary civil rights and freedoms; the total control over millions of workers in Asian sweatshops, which are often organized as concentration camps; the massive use of forced labour in the exploitation of natural resources in many central African states (for example the Congo).

Nowhere is this reversal of virtuality into materiality more brutally displayed than in the approaching end of nature. Reading and watching reports on the devastating effect of Hurricane Irma, I was reminded of Trisolaris, a strange planet from *The Three-Body Problem*, Liu Cixin's sci-fi masterpiece. A scientist is drawn into a virtual-reality game, 'Three Body', in which players find themselves on the alien planet Trisolaris, whose three suns rise and set at strange and unpredictable intervals and positions: sometimes too far away, making the planet horribly cold; sometimes too close and destructively hot; and sometimes not at all for long periods of time. The players can somehow dehydrate themselves and the rest of the population to weather the worst seasons, but life is a constant struggle against apparently unpredictable elements. Despite that, players slowly find ways to build civilizations and attempt to predict the strange cycles of heat and cold. Do phenomena like Irma not demonstrate that our Earth itself is gradually turning into Trisolaris? Devastating hurricanes, droughts and floods, not to mention global warming – do they all not indicate that we are witnessing something for which the only appropriate term is 'the end of nature'? 'Nature' is to be understood here in the traditional sense of a regular rhythm of seasons, the reliable background of human history, something which we can count on always being there.

It is difficult for an outsider to imagine how it feels when a vast domain of densely populated land disappears under water, and millions are deprived of the very basic coordinates of their life-world: the land with its fields, but also with its cultural monuments, the stuff of their dreams, is no longer there, so that, although surrounded by water, they are in a way like fish out of water – it is as if the environs that thousands of generations have taken as the foundation of their lives starts to crack. Similar catastrophes have, of course, been known for centuries, some even from the very prehistory of humanity. What is new today is that, since we live in a 'disenchanted' post-religious era, such catastrophes can no longer be rendered meaningful as part of a larger natural cycle or as an expression of divine wrath. Consider how, back in 1906, the American philosopher William James described his reaction to an earthquake:

> emotion consisted wholly of glee, and admiration. Glee at the vividness which such an abstract idea as 'earthquake' could take on when verified concretely and translated into sensible reality . . . and admiration at the way in which the frail little wooden house could hold itself together in spite of such a shaking. I felt no trace whatever of fear; it was pure delight and welcome.[15]

How far are we here from the shattering of the very foundations of our life-world!

Phenomena like global warming make us aware that, with all the universality of our theoretical and practical activity, we are at a certain basic level just another species living on the planet Earth. Our survival depends on certain natural parameters that we take for granted. The lesson of global warming is that the freedom of humankind is possible only against the background of a stable environment (temperature, the composition of the air, sufficient water and energy supplies, and so on): humans can 'do what they want' only insofar as they remain marginal enough, so that they don't seriously perturb the parameters of life on earth. The limitation of our freedom that becomes palpable with global warming is the paradoxical outcome of the very exponential growth of our freedom and power, i.e. of our growing ability to transform nature around us, even to the point of destabilizing the basic geological foundations of life. 'Nature' thereby literally becomes a socio-historical category, but not in the exalted young-Lukács sense, in which the content of what is for us (counts for us as) 'nature' is always overdetermined by a historically specified social totality that structures the transcendental horizon of our understanding of nature. Rather, it becomes a socio-historical category in the much more radical and literal (ontic) sense of something that is not just a stable background to human activity, but is affected by it in its very basic components.

We are thus entering a new phase in which it is simply nature itself that melts into air: the main consequence of the scientific breakthroughs in biogenetics is the end of nature. Once we know the rules of their construction, natural organisms are transformed into objects amenable to manipulation. Nature, human and inhuman, is thus 'desubstantialized', deprived of its impenetrable density, of what Heidegger called 'earth'. This compels us to give a new twist to the title of Freud's *Unbehagen in der Kultur* – discontent, uneasiness, in culture. The title is usually translated as 'civilization and its discontents', thus missing the opportunity to bring into play the opposition of culture and civilization: discontent is *in* culture, its violent break with nature, while civilization can be conceived as precisely the secondary attempt to patch things up, to 'civilize' the cut, to reintroduce the lost balance and an appearance of harmony. With the latest developments, the discontent shifts from culture to nature itself: nature is no longer 'natural', the reliable 'dense' background of our lives; it now appears as a fragile mechanism which, at any point, can explode in a catastrophic direction.

Nature is in increasing disorder, not because it overwhelms our cognitive capacities but primarily because we are not able to master the effects of our own interventions in its course – who knows what the ultimate consequences of our biogenetic engineering or of global warming will be? The surprise comes from us, it concerns the opacity of our role: the problem is not some cosmic mystery like the explosion of a supernova, it is us ourselves, our collective activity. This is what we call 'anthropocene': a new epoch in the life of our planet in which we humans can no longer rely on the Earth as a reservoir ready to absorb the consequences of our productive activity. We must acknowledge that we live on a 'Spaceship Earth', and be responsible and accountable for its condition. At the very moment at which we become powerful enough to affect the most basic elements of our life, we have to accept that we are just another animal species on a small planet. A new way to relate to our environs is necessary once we realize this: we must become modest agents collaborating with our environment, permanently negotiating a tolerable level of stability, with no *a priori* formula to guarantee our safety.

Does this mean that we should assume a defensive approach and search for a new limit, a return to (or, rather, the invention of) some new balance? This is what the predominant ecological thinking proposes we should do, and the same goal is pursued by bioethics with regard to biotechnology: biotechnology explores new possibilities of scientific intervention (genetic manipulations, cloning and so on), and bioethics endeavours to impose moral limitations on what biotechnology enables us to do. As such, bioethics is not immanent in scientific practice: it intervenes in this practice from outside, imposing external morality on it. One can even say that bioethics is the betrayal of the ethics inherent in scientific endeavour, the ethics of 'do not compromise your scientific desire, follow its path inexorably'. Such attempts at limitation fail because they ignore the fact that there is no objective limit: we are discovering that the object itself – nature – is not stable. Sceptics like to point out the limits of our knowledge of nature; however, these limits in no way imply that we should underplay the ecological threat. On the contrary, we should be even more careful about it, since the situation is profoundly unpredictable. The recent uncertainties about global warming do not signal that things are not too serious, rather that they are even more chaotic than we thought, and that natural and social factors are inextricably linked.

Can we, then, use capitalism itself against this threat? Although capitalism can easily turn ecology into a new field of capitalist investment and competition, the very nature of the risk involved fundamentally precludes a market solution – why? Capitalism only works in precise social conditions: it implies trust in the objectivized mechanism of the market's 'invisible hand' which, as a kind of cunning of reason, guarantees that the competition of individual egotisms works for the common good. However, we are in the midst of a radical change: what looms on the horizon today is the unheard-of possibility that a subjective intervention will trigger an ecological catastrophe, a fateful biogenetic mutation, a nuclear or similar military-social disaster. For the first time in human history, the act of a single socio-political agent can alter and interrupt the global historical and even natural process.

Jean-Pierre Dupuy refers here to the theory of complex systems, which accounts for their two opposing features: their robust, stable character and their extreme vulnerability. These systems can accommodate themselves to great disturbances, integrate and find new balance and stability – up to a certain threshold (a 'tipping point'), above which a small disturbance can cause a total catastrophe and lead to the establishment of a completely different order. For centuries, humanity did not have to worry about the impact on the environment of its productive activity – nature was able to accommodate itself to deforestation, to the use of coal and oil, and so on. However, we cannot be sure if, today, we are not approaching a tipping point – we really cannot be sure, since such points are only clearly perceived once it is already too late, *in retrospect*. Either we take the threat of environmental catastrophe seriously and decide today to do things that, if the catastrophe doesn't occur, will appear ridiculous, or we do nothing and lose everything if it does happen. The worst scenario is to choose a middle ground, to take a limited number of measures – in this case, we will fail whatever occurs (that is to say, the problem is that there is no middle ground: either the environmental catastrophe will occur or it won't). In such a situation, the talk about anticipation, precaution and risk-control becomes meaningless.

The main lesson to be learned is, therefore, that humankind should be prepared to live in a more 'plastic' and nomadic way: local or global changes in the environment may need unheard-of large-scale social transformations. Let us say that a new, gigantic volcanic eruption makes the whole of Iceland uninhabitable: where will the people of Iceland move? Under what conditions? Should they be given a piece of land or just be dispersed around the world? What if northern Siberia becomes more inhabitable and suitable for agriculture, while large Sub-Saharan regions become too dry for a large population to live there – how will the exchange of populations be organized? When similar things happened in the past, social changes occurred in a spontaneous way, with violence and devastation; such a prospect is catastrophic in today's conditions, with weapons of mass-destruction so readily available. One thing is clear: national sovereignty will have to be radically redefined and new levels of global cooperation invented. And what about the immense changes in the economy and consumption that will be needed as a result of new weather patterns or shortages of water and energy sources? Through what decision-making processes will such changes be executed?

And, last but not least, it is important to bear in mind the strange coincidence of opposites in the threats that confront us: trouble comes from 'material' outside (the end of nature, environmental catastrophes) and from the inside, from the 'immaterial' virtual sphere (who controls the digital space that controls us? Who manipulates hackers?).

Of Mice and Men, or Towards Posthuman Capitalism

'Nosedive', the first episode of the third series of *Black Mirror*, is set in an alternative reality where people can rate one another using their phones, and where your ratings can impact on your entire life. It tells the story of Lacie, a young woman overly obsessed with her ratings who, after being chosen by her popular childhood friend as maid of honour at her wedding, sees it as an opportunity to improve her ratings and achieve her dreams – access to many places is allowed only with a rating above 4.5 (out of 5). She fails and everything goes wrong, and eventually she has a ranking of 0; she then has the technology to be rank-removed and is jailed for her actions. While in her cell, Lacie begins to exchange insults with another prisoner, and their mutual anger turns to mutual delight as they realize they are free to do it. But is this an alternative reality? According to a report in *Business Insider*, 'China might use data to create a score for each citizen based on how trustworthy they are':

> The Chinese government is planning on implementing a system that connects citizens' financial, social, political, and legal credit ratings into one big social-trustability score. The idea would be that if someone breaks trust in one area, they'd be adversely affected everywhere. The Chinese plan for a more widespread scoring system has been in the works since 2015. But in September, the government released bullet points of proposed penalties for those who 'breaks social trust' (which could be done by defaulting on a loan, for example, or voicing a dissenting opinion against the government online). According to

> the policy documents, here's what could happen if you're a low scorer: You won't be considered for public office, You'll lose access to social security and welfare, You'll be frisked more thoroughly when passing through Chinese customs, You'll be shut out of senior level positions in the food and drug sector, You won't get a bed in overnight trains, You'll be shut out of higher-starred hotels and restaurants and will be rejected by travel agents, Your children won't be allowed into more expensive private schools.[16]

You may think this is really only another story about Chinese totalitarianism. But are we not already doing the same, just in a more discreet way? Instead of looking at how data is gathered when we apply for a job or ask for bank credit, let's examine a more subtle example:

> At a recent Transport for London talk the possibility of 'gamifying' commuting within London was discussed. In order to facilitate this possibility TfL have made the internet API and data streams used to monitor all London Transport vehicles (buses, tube trains, overground trains, ferries) Open Source and Open Access in the hope that app developers will build London-focused apps based around the public-transport system, maximizing profit. One idea is that if a particular tube station is becoming clogged up due to other delays, TfL could give 'in-game rewards' for people willing to use alternative routes and thus smooth out the jam. Whilst traffic-jam prevention may not seem like evidence of a dystopia of total corporate and state control, it actually shows the dangerous potentiality in such technologies. It shows that the UK is not so far away from the 'social-credit' game system planned for implementation in Beijing, to rate each citizen's trustworthiness and give them rewards for their dedication to the Chinese state. Whilst the UK mainstream media reacted with shock to these innovations in Chinese app development, a closer look at the current electronic structures of mapping and controlling our movements shows that a similar framework is already in its development phase in London too. In the 'smart city' to come it won't be just traffic jams that are smoothed out but any inefficient misuse or dangerous occupation of space.[17]

Furthermore, one should bear in mind that such grading is never all-encompassing: it always presupposes a double exemption. Decades ago, *Mad* magazine published a series of variations on the theme of four levels of hierarchy. With regard to fashion, say, there are those at the bottom who live outside fashion and simply don't care about it; then there are those who try to catch up with fashion but always lag behind; then those who can afford to be fully in sync with the latest trends; and, finally, those at the very top who, like those at the bottom, don't care what they wear because they determine fashion – what they decide to wear *is* the fashion. Will it not be the same with social trust? At the bottom are the outsiders who don't care about their grading; then there are those who lag behind and try to elevate their grades; then those who achieve top grades; and finally, again, those who, like those at the bottom, don't care about their grading because everything is accessible to them (in China, for example, top members of the state *nomenklatura* will certainly not have to worry about their grades). The top and the bottom groups are both in a sense free: they don't worry about their grading, and it can even be said that those at the bottom are more free since those at the top have other worries (will they stay at the top?). Perhaps those at the bottom, excluded as they are from grading and proudly ignoring it, are today's new proletarians who are, as Marx pointed out, free in the double sense – free in the sense of having no social possessions, and simply being free.

Today, those above grading are, of course, great corporations linked to government agencies – they exemplify the privatization of our commons. The figure of Elon Musk is emblematic here – he belongs to the same group as Bill Gates, Jeff Bezos, Mark Zuckerberg etc., all 'socially conscious' billionaires. They stand for global capital at its most seductive and 'progressive' – in short, at its most dangerous. Musk likes to warn about the threats the new technologies pose to human dignity and freedom – which, of course, doesn't prevent him from investing in a brain-computer interface venture called Neuralink, a company which is focused on creating devices that can be implanted in the human brain, with the eventual purpose of helping human beings to merge with software and keep pace with advances in artificial intelligence. These advances could improve memory or allow for more direct interfacing with computing devices: 'Over time I think we will probably see a closer merger of biological intelligence and digital intelligence.'[18] Every technological innovation is first presented like this, its health or humanitarian benefits emphasized, which blinds us to more ominous implications and consequences: can we even imagine what new forms of control this so-called 'neural lace' contains? This is why it is absolutely imperative to keep it out of the control of private capital and state power and to render it totally accessible to public debate. Julian Assange was right in his strangely ignored key book on Google:[19] to understand how our lives are regulated today, and how this regulation is experienced as our freedom, we have to focus on the shadowy relation between the private corporations that control our commons and the secret state agencies.

This is what makes Assange such a threat to the establishment, and one can only imagine the behind-the-scenes pressures recently exerted by Western powers on Ecuador so that the small country could not but add another turn of the screw to the isolation of Julian Assange from public space: his internet access is now cut off, many visitors are refused entrance . . . a slow social death of a person who has for almost six years been confined to an apartment in the Ecuadorian embassy in London. It happened for a short period once before, at the time of the US elections – but at that time it was a reaction to WikiLeaks publishing documents that could affect the outcome of those elections, while there is no such excuse now: Assange's 'meddling' in international relations consisted only of publishing on the web his opinions about the Catalonia crisis and the Skripal poisoning scandal. So why such action now, and why did it cause so little uproar among the public?

Regarding that second question, it is not enough to claim that people have simply got tired of Assange: a key role has been played by the long, slow and well-orchestrated campaign of character assassination that reached the lowest level imaginable a few months ago with unverified rumours according to which the Ecuadorians wanted to get rid of him because of his bad smell and dirty clothes. In the first stage of these personal attacks, his ex-friends and collaborators went public to say that WikiLeaks began well but then got bogged down with Assange's political bias (his anti-Hillary obsession, his suspicious ties with Russia . . .). This was followed by more direct personal defamations: he is paranoiac and arrogant, obsessed by power and control . . . Now we have reached the direct bodily level of smells and stains.

Assange a paranoiac? When you live permanently in an apartment that is bugged from above and below, a victim of constant surveillance organized by the secret services, who wouldn't be that? Megalomaniac? When the (now ex-)head of the CIA says your arrest is his priority, does not this imply that you are a 'big' threat to some, at least? Behaving like the head of a spy organization? But WikiLeaks *is* a spy organization, although one that serves the people, keeping them informed about what goes on behind the scenes. Is Assange a refugee from justice, hiding in the Ecuadorian embassy to escape judgement? But what kind of justice is it that threatens to arrest him when the charges are dropped?

So let's move to the big question: why now? I think one name explains it all: Cambridge Analytica – a name that symbolizes what Assange is all about, what he fights for: the disclosure of the links between the great private corporations and government gencies. Remember what a big topic and obsession the Russian meddling in the US elections was – now we know that it was not Russian hackers (supposedly working with Assange) who nudged the people towards Trump but our own data-processing agencies, who allied their forces with political ones. This doesn't mean that Russia and their allies are innocent: they probably did try to influence the outcome, in the same way that the US does in other countries (only, in that case, it is called 'helping democracy'). But it means the big bad wolf who distorts our democracy is right here, not in the Kremlin – and this is what Assange was claiming all the time.

But where, exactly, is this big bad wolf? To grasp the whole scope of this control and manipulation one should move beyond the link between private corporations and political parties (as is the case with Cambridge Analytica), to the interpenetration of data-processing companies like Google or Facebook and state security agencies. We shouldn't be shocked at China, but rather at ourselves, when we accept the same regulation while believing that we retain our full freedom and that our media just help us to realize our goals (at least in China people are fully aware that they are regulated). The overall image emerging, combined with what we also know about the latest developments in biogenetics, provides an adequate and terrifying vision of new forms of social control that make the good old twentieth-century 'totalitarianism' a rather primitive and clumsy machine of control.

The biggest achievement of the new cognitive-military complex is that direct and obvious oppression is no longer necessary: individuals are much better controlled and 'nudged' in the desired direction when they continue to experience themselves as free and autonomous agents of their own life. This is another key lesson of WikiLeaks: our unfreedom is most dangerous when it is experienced as the very medium of our freedom – what can be more free that the incessant flow of communications that allows every individual to popularize her opinions and form virtual communities at her own free will? This is why it is absolutely imperative to keep the digital network out of the control of private capital and state power, i.e. to render it totally accessible to public debate.

Now we can see why Assange has to be silenced when the subject of Cambridge Analytica is all over our media. Those in power attempt to reduce the issue to a particular 'misuse' by some private corporations and political parties – but where is the state itself, the half-invisible apparatuses of the so-called 'deep state'? No wonder that the *Guardian*, which has extensively reported on the Cambridge Analytica 'scandal', recently published a disgusting attack on Assange as a megalomaniac and fugitive from justice. Write as much as you want about Cambridge Analytica and Steve Bannon, just don't dwell on what Assange was drawing our attention to: that the state apparatuses that are now expected to investigate the 'scandal' are themselves part of the problem.

Assange characterized himself as the spy of and for the people: he is not spying on the people for those in power, he is spying on those in power for the people. This is why the only ones who can really help him now are we, the people. Only our pressure and mobilization can alleviate his predicament. One often reads how the Soviet secret service not only punished its traitors even if it took decades to do it, but

*

also fought doggedly to free them when they were caught by the enemy. Assange has no state behind him, just us, the people – let us do at least what the Soviet secret service was doing, let's fight for him no matter how long it will take!

Digital regulation has lately taken a much more ominous turn. John Steinbeck titled his famous novella after Robert Burns's poem 'To a Mouse':

> I'm truly sorry Man's dominion
> Has broken Nature's social union,
> And justifies that ill opinion,
> Which makes thee startle,
> At me, thy poor, earth-born companion,
> And fellow-mortal!
> [. . .]
> The best laid schemes of Mice and Men
> Gang aft agley,*
> And leave us nought but grief and pain,
> For promised joy!

The situation described in these lines is that of a human being apologizing to a mouse that he has broken 'nature's social union' due to his thirst for domination over the natural world, which justifies the mouse's ill opinion and fear of humans. The human also concedes that, even if his scheme was well meant, it turned ugly, causing nothing but grief and pain. Can we imagine the same scene taking place between the human scientist and the mouse on whom he performs the following experiment?

> Imagine someone remotely controlling your brain, forcing your body's central processing organ to send messages to your muscles that you didn't authorize. It's an incredibly scary thought, but scientists have managed to accomplish this science-fiction nightmare for real, albeit on a much smaller scale, and they were even able to prompt their test subjects to run, freeze in place, or even completely lose control over their

* 'Agley' is a Scottish word for 'awry', or 'wrong'.

> limbs. Thankfully, the research will be used for good rather than evil . . . for now. The effort, led by physics professor Arnd Pralle, PhD, of the University at Buffalo College of Arts and Sciences, focused on a technique called 'magneto-thermal stimulation'. It's not exactly a simple process – it requires the implantation of specially built DNA strands and nanoparticles which attach to specific neurons – but once the minimally invasive procedure is over, the brain can be remotely controlled via an alternating magnetic field. When those magnetic inputs are applied, the particles heat up, causing the neurons to fire . . . Despite only being tested on mice, the research could have far-reaching implications in the realm of brain research. The holy grail for dreamers like Elon Musk is that we'll one day be able to tweak our brains to eliminate mood disorders and make us more perfect creatures. This groundbreaking research could very well be an important step towards that future.[20]

The qualified hope that this research 'will be used for good rather than evil . . . for now' sounds like the familiar doctor jokes about 'first the good news, then the bad news'. When a new invention like the direct digitalization of our brain is sold to the public, the media generally begin by pointing out its medical benefits and new opportunities to diminish suffering. Even the renowned Stephen Hawking's little finger – the minimal link between his mind and outside reality, the only part of his paralysed body that he was able move – would no longer be needed: with a wired mind, he would have been able to make his wheelchair move, i.e. his brain could have served as a remote-control machine. But, as they say, what goes around comes around – the digitalization of our brains opens up unheard-of new possibilities for control. Incidentally, the news I quoted is no longer really news: even in May 2002 it was reported that scientists at New York University had attached to a rat's brain a computer chip able to receive signals, so that one could control the rat, determining the direction in which it runs, by means of a steering mechanism, as with a remote-controlled toy car. For the first time, the 'will' of a living animal agent, its 'spontaneous' decisions about the movements it makes, were taken over by an external machine. Of course, the big philosophical question here is: how did the unfortunate rat 'experience' its movement, which was effectively decided from outside? Did it continue to experience it as something spontaneous (i.e., was it totally unaware that it was being steered), or did it realize that 'something was wrong', that an external power was deciding its movements?

Even more crucially, let us apply the same reasoning to an identical experiment performed on humans (which, ethical questions notwithstanding, shouldn't be much more complicated, technically speaking, than those on a rat). One can argue that one should not apply the human category of 'experience' to the rat, but we must ask the same question of a human being. So, again, will I, as a steered human being, continue to experience my movements as something spontaneous? Will I remain totally unaware that my movements are being steered, or will I realize that something is wrong, that an external power is determining them? And, how, precisely, will this 'external power' manifest itself – as something inside me, an unstoppable inner drive, or as a simple external coercion? If I remain totally unaware that my 'spontaneous' behaviour is steered from outside, can we really go on pretending that this has no consequences for our notion of free will?

With a little bit of irony, we can already identify such a steered human being in our political reality: when Alexis Tsipras, a partisan of anti-austerity politics, having triumphantly won the referendum saying 'no' to EU financial pressure, suddenly changed his position and agreed to enact the toughest austerity politics, was it not as if the financial-political powers in Brussels had pressed a button and made him act like their remote-controlled toy? Many observers have noted that, after this change, when Tsipras appears on TV in the company of big European leaders, there is something strange in his behaviour: he often just stands and smiles, as if he is not fully aware of what he is doing. However, the picture is more complex. On 17 October 2017 Tsipras visited Trump in the White House and, speaking in the Rose Garden after their meeting, he proclaimed that he was now ready to go into partnership with Trump: 'We have common values. Don't forget that the value of democracy and freedom was born in Greece. It's one of the values that traverses American culture and American tradition. The President now continues that tradition.'[21] The Tsipras we see here is not the puppet of an EU imposing tough austerity politics – we discern a kind of authentic personal surplus, no international capital is pressuring him. His praise of Trump is decidedly more subjective, and, geopolitically, it makes sense (the US, annoyed with Turkey, counts on Greece again), in the same vein as his declaration, during his visit to Israel a couple of years ago, that Jerusalem is the eternal capital of the Jews, or his fraternizing with Serbia and other Orthodox countries. One can also understand that, after bitter experiences with the EU, he is looking for support from those who are also basically anti-European. However, none of this in any way justifies his position: we are simply dealing with the tragi-comic consequences of his capitulation to EU blackmail.

While this image of Tsipras as a remote-controlled toy is of course no more than a political joke in rather bad taste, big questions emerge here – not only basic philosophical ones but also political ones: when Musk says, 'we'll one day be able', who will this 'we' be? Corporations, government, anyone with money? One thing is clear: science and philosophy will have to combine their forces. It happens from time to time that a similar idea appears in two different fields of theory that do not intercommunicate at all – say, in 'postmodern' speculation and in empirical science. This is what has occurred over the last few decades with the idea of theoretical anti-humanism or of an inhuman subject, which played an important role in contemporary French thought, from Foucault and Lacan to Badiou. Lately, cognitive sciences have proposed their own version of anti-humanism: with the digitalization of our lives and the prospect of a direct link between our brain and digital machinery, we are entering a new posthuman era in which our basic self-understanding as free and responsible human agents will be affected. In this way, posthumanism is no longer an eccentric theoretical proposal but a matter concerning our daily lives. Can these two aspects be brought together into a unique theoretical perspective, or are they condemned to speak different languages ('postmodern' theory reproaching cognitivism as a naive naturalist determinism, and cognitivists dismissing 'postmodern' theory as an irrelevant speculation that remains rooted in traditional philosophical space)?[22] The first thing to note here is how the rise of posthuman agents and the anthropocene epoch are two aspects of the same phenomenon: at exactly the time when humanity becomes the main geological factor threatening the entire balance of the life of Earth, it begins to lose its basic features and transforms itself into posthumanity.

The question that underlies this problem is: how are capitalism and the prospect of posthumanity related? Usually it is posited that capitalism is more historical and our humanity, including our sexual differences, more basic, even ahistorical; however, what we are witnessing today is nothing less than an attempt to integrate the passage to posthumanity with capitalism. This is what the efforts of new billionaire gurus like Elon Musk are all about: their predictions that capitalism 'as we know it' is coming to an end refer to 'human' capitalism, and the passage they talk about is the passage from human to posthuman capitalism. *Blade Runner 2049* deals with this topic – here is the storyline (again, shamelessly taken from Wikipedia):[23]

> In 2049, replicants (bioengineered humans) have been integrated into society as servants and slaves. K, a newer replicant model created to obey, works as a 'blade runner' for the LAPD, hunting down and 'retiring' rogue older model replicants. His home life is spent with his holographic girlfriend Joi, an artifical-intelligence product of the Wallace Corporation. [One should note that this Wallace Corporation, which produces posthuman replicants, grew out of the crisis of the anthropocene: it gained its power by saving humanity from the hunger that resulted from a collapse of natural reproduction due to excessive human activity – it found a way to produce huge quantities of artificial food.] K's investigation into a growing replicant freedom movement leads him to a farm, where he retires rogue replicant Sapper Morton and finds a buried box. Forensic analysis reveals the box contains the remains of a female replicant who died as the result of complications from an emergency caesarian section. K finds this unsettling, as pregnancy in replicants was originally thought to be impossible.
>
> K is ordered to destroy all evidence related to the case and to retire the child by his superior, Lieutenant Joshi, who believes the knowledge that replicants are able to reproduce to be dangerous and could lead to war, since it blurs the clear line of separation between replicants and humans. K, disturbed by his orders to kill a born individual, visits the headquarters of Wallace Corporation and meets its founder, Niander Wallace, who identifies the body as Rachael, an experimental replicant. In the process, he learns of her romantic ties with former veteran blade runner Rick Deckard. Believing that reproduction in replicants can bolster his production and expand his off-world operations, but lacking

the technology to give them this ability himself, Wallace sends his replicant enforcer Luv to steal Rachael's remains from LAPD headquarters and follow K to find Rachael's child.

Returning to Morton's farm, K finds a hidden date that matches a childhood memory about hiding a toy horse, which he later finds at an orphanage, suggesting that his memories – which he thought were implants – are real; Joi insists this is evidence that K is in fact a real person. While searching birth records for that year, he discovers that twins were born on that day with identical DNA except for the sex chromosome; only the boy is listed as alive. K seeks out Dr Ana Stelline, a memory designer for Wallace Corporation, who informs him that it is illegal to program replicants with humans' real memories, leading K to believe he might be Rachael's son. After failing a test of his replicant behaviour, K is suspended by Joshi, but Joshi gives him forty-eight hours to disappear. After transferring Joi to a mobile emitter, despite knowing that if it is damaged she will be erased, K has the toy horse analysed and finds traces of radiation that lead him to the ruins of Las Vegas, where he finds Deckard. Deckard reveals that he scrambled the birth records to cover his tracks and was forced to leave a pregnant Rachael with the replicant freedom movement to protect her.

Luv and her men murder Joshi, track K's location and arrive to kidnap Deckard. They leave a badly injured K for dead and destroy Joi's emitter. He is later rescued by the replicant freedom movement, who were also tracking him. He is told by their leader, Freysa, that he is wrong to think he has a unique role to fulfil in the movement, and that Rachael's child is actually a girl. K deduces that Stelline is Deckard's daughter, as she is the only one capable of creating the memory and implanting it into him. Freysa urges K to prevent Wallace from uncovering the secrets of replicant reproduction by any means necessary, including killing Deckard.

It is important that this part of the film takes place in Las Vegas. Many reviewers have noted Tarkovskian echoes in the film, not only in its slow rhythms but also in its landscape, which evokes the Zone from Tarkovsky's *Stalker*. In the first *Blade Runner*, the megalopolis itself is a Zone, while outside the Zone there is unspoiled green nature (to which Deckard and Rachael escape at the film's end, at least in the first released version); in *Blade Runner 2049*, Earth's entire surface is a poisonous Zone (the story takes place after a global ecological catastrophe) – but there is a kind of Zone within a Zone: the area around Las Vegas, where Deckard is hiding, the irradiated territory where only replicants can survive, and where the human police force can only intervene briefly, heavily protected with masks. In the Las Vegas Zone time is stuck in an eternal circular movement which is perfectly rendered by the eternal self-replicating hologram show of old stars (Elvis Presley, etc.) on an abandoned stage. There is nothing glamorous about it, the hologram hyper-reality is fractured, interrupted again and again; but one should nonetheless bear in mind that here we are in a kind of liberated territory, or at least a territory abandoned by state power where the replicants' resistance is based too.

Back in Los Angeles, Deckard is brought before Wallace, who suggests that Rachael's feelings for him were engineered by Tyrell to test the possibility of a replicant becoming pregnant. When Deckard refuses to cooperate, Wallace has Luv escort him to off-world outposts to be tortured for information, but K intercepts them, killing Luv and staging Deckard's death to protect him from both Wallace and the replicants. He leads Deckard to Stelline's office and laments that his best memories belong to her. Deckard cautiously enters the office and approaches Stelline, while K succumbs to his wounds.

So why is the fact that two replicants (Deckard and Rachael) formed a sexual couple and created a human being in a human way experienced as such a traumatic event, celebrated by some as a miracle and viewed with horror by others as a threat? Is it about reproduction or about sex, that is, about sexuality in its specific human form? The movie focuses exclusively on reproduction, again neglecting the big question: can sexuality, deprived of its reproductive function, survive into the posthuman era?

The image of sexuality remains the standard one: the sexual act is shown from the male perspective, so that the flesh-and-blood android woman is reduced to being a physical prop for the hologram fantasy woman Joi, created to serve the man: 'she must overlap with an actual person's body, so she is constantly slipping between the two identities, showing that the woman is the real divided subject, and the flesh-and-blood Other just serves as a vehicle for the fantasy'.[24] Joi is not materialized but is a programmed male fantasy, made to jump from being home companion to housemaid and to sex worker adapted to her owner's wishes and desires. The sex scene in the film is thus almost too directly 'Lacanian' (in line with films like *Her*), ignoring authentic heterosexuality in which the partner is not just a prop for me to enact my fantasies but a real Other.[25] The movie also fails to explore the potentially antagonistic difference among androids themselves, between the 'real-flesh' androids and those whose bodies are just 3-D hologram projections: how does, in the sex scene, the flesh-and-blood android woman relate to being reduced to the physical prop for the male fantasy? Why doesn't she resist and sabotage it?

The movie provides a whole panoply of modes of exploitation, including a half-illegal entrepreneur using child labour (hundreds of human orphans) to scavenge old digital machinery. From a traditional Marxist standpoint, strange questions arise here: if fabricated androids work, is exploitation still operative, does their work produce value that is in excess of their own value as commodities so that it can be appropriated by their owners as surplus value? One should note that the idea of enhancing human capacities in order to create perfect posthuman workers or soldiers has a long history in the twentieth century. In the late 1920s, none other than Stalin for a while financially supported the 'human ape' project proposed by the biologist Ilya Ivanov (a follower of Alexander Bogdanov, the target of Lenin's critique in *Materialism and Empirio-Criticism*): the idea was that by coupling humans and orangutans one could create a perfect worker and soldier impervious to pain, tiredness and bad food. In his innate racism and sexism, Ivanov, of course, tried to couple male humans and female apes; plus the humans he used were black males from the Congo, since they were supposed to be genetically closer to apes – the Soviet state financed an expensive expedition there. When his experiments failed, Ivanov was liquidated. The Nazis also regularly used drugs to enhance the fitness of their elite soldiers, and the US Army is now experimenting with genetic changes and drugs to make soldiers super-resilient (for example, they already have pilots ready to fly and fight for seventy-two hours).

In the domain of fiction, one should include zombies in this list. Horror movies register class difference in the demeanour of vampires and zombies: vampires are well mannered, exquisite, aristocratic, living among normal people, while zombies are clumsy, inert, dirty and attack from the outside, like a primitive revolt of the excluded. The equation between zombies and the working class was directly made in *White Zombie* (1932, Victor Halperin), the first pre-Hays Code full-length zombie film. There are no vampires in this film – but, significantly, the main villain who controls the zombies is played by Bela Lugosi, who had become famous the previous year as Dracula. *White Zombie* takes place on a plantation in Haiti, the site of the most famous slave revolt. Lugosi receives another plantation owner and shows him his sugar factory, where workers are zombies who, as Lugosi is quick to explain, don't complain about long working hours, demand no trade unions, never strike, but just go on and on working. A film like this was only possible before the imposition of the Hays Code.

In yet another nice reversal of the standard formula in which the hero, living as (and regarding himself as) just an ordinary guy, discovers he is an exceptional figure with a special mission, K in *Blade Runner 2049* thinks he is the special figure everybody is looking for (the child of Deckard and Rachael), but gradually realizes that (like many other replicants) he is just an ordinary replicant obsessed with an illusion of greatness, so he ends up sacrificing himself for Stelline, the truly exceptional figure. The enigmatic Stelline is crucial here: she is the 'real' (human) daughter of Deckard and Rachael, which means that she is the human daughter of replicants, thus inverting the process of man-made replicants. Living in her isolated world (unable to survive in open spaces filled with real plants and animal life), condemned to utter sterility (white clothes in an empty room with white walls), her contact with life limited to the virtual universe generated by digital machines, she is ideally positioned as a creator of dreams (she works as an independent contractor, programming false memories to be implanted into replicants). As such, Stelline is in no way a subversive agent – she is a freelance who works for Wallace and, as such, directly collaborates in the prevention of the replicants' ability to rebel – in short, she is the chief ideologist of the Wallace corporation, mass-producing dreams and memories to keep the subjects satisfied. Would the resistance movement not be fully justified in kidnapping her and making her work for them? Stelline exemplifies the absence (or, rather, impossibility) of sexual relationships, which she supplants with a rich fantasmatic tapestry. No wonder that the couple created at the film's end is not the standard sexual one but the asexual pairing of father and daughter. This is why the final shots of the film are so familiar and weird at the same time: K sacrifices himself in a Christ-like gesture on the snow to create the father-daughter couple.

Is there a redemptive power in this reunion? Or should we read its significance against the background of the film's symptomatic silence about social friction among humans in the society it depicts: where do human 'lower classes' stand? However, the film does render nicely the antagonism that cuts across the ruling elite itself in global capitalism: that between the state and its apparatuses (personified in Joshi) and big corporations (personified in Wallace) pursuing progress to its self-destructive end:

> While the state political-legal position of the LAPD is one of potential conflict, Wallace sees only the revolutionary productive potentials of self-reproducing replicants, which he hopes could give him a leg up in his business. His perspective is one of the market; and it is worth looking at these contradictory perspectives of Joshi and Wallace, for it is indicative of the contradictions that do exist between the political and the economic; or, put differently, it oddly indicates the intersection of the class state mechanism and the tensions in the economic mode of production.[26]

Although Wallace is a real human, he acts as inhuman, an android blinded by excessive desire, while Joshi stands for apartheid, for the strict separation of humans and replicants – or, to quote her: 'There is an order to things. The world is built on a wall that separates kind. Tell either side there's no wall, you've bought a war . . . or a slaughter.' Her view is that if this separation is not upheld, there is war and disintegration. 'If a child is born from a replicant mother (or parents), does he remain a replicant? If he has produced his own memories, is he still a replicant? What is now the dividing line between humans and replicants if the latter can self-reproduce? What marks our humanity?'[27] So should we not, mindful of *Blade Runner 2049*, supplement the famous passage in *The Communist Manifesto*, adding that sexual 'one-sidedness and narrow-mindedness become more and more impossible', that in the domain of sexual practices 'all that is solid melts into air, all that is holy is profaned', and that capitalism tends to replace the standard normative heterosexuality with a proliferation of unstable shifting identities and/or orientations? Today's celebration of 'minorities' and 'marginals' *is* the predominant majority position: even alt-rightists who complain about the terror of liberal political correctness present themselves as protectors of endangered minorities. Or take those critics of patriarchy who attack it as if it were still a hegemonic position, ignoring what Marx and Engels wrote more than 150 years ago in the first chapter of *The Communist Manifesto*: 'The bourgeoisie, wherever it has got the upper hand, has put an end to all feudal, *patriarchal*, idyllic relations.' This is still ignored by those Leftist cultural theorists who focus their critique on patriarchal ideology and practice. Not to mention the prospect of new forms of android (genetically or biochemically manipulated) posthumanity which will shatter the very separation of human and inhuman.

So why do the new generation of replicants not rebel?

> Unlike the replicants in the original, the newer replicants never revolt, though it is not clearly explained why, other than they are programmed not to. The film, however, hints at the explanation: the fundamental difference between the new and old replicants involves their relation to their false memories. The older replicants revolted because they believed their memories to be real and thus could experience the alienation of recognizing that they weren't. The new replicants know from the beginning that their memories are faked, so they are never deceived. The point is thus that fetishistic disavowal of ideology renders subjects more enslaved to the ideology than simple ignorance of its functioning.[28]

The new generation of replicants are deprived of the illusion of authentic memories, of all substantial content of their being, and are thereby reduced to the void of subjectivity, i.e. to the pure proletarian status of *substanzlose Subjektivität*. So does the fact that they don't rebel mean that rebellion has to be sustained by the awareness of some authentic content threatened by the oppressive power?

K stages a fake accident to make Deckard disappear, not only from the sight of the state and capital (the Wallace Corporation) but also from that of the replicant rebels (who are led by a woman, Freysa, a name which, of course, echoes freedom, *Freiheit*, in German).

Although one can justify his decision by the fact that Freysa also wants Deckard dead so that Wallace will not be able to discover the secret of replicant reproduction – both the state apparatus (embodied in Joshi) and revolutionaries (embodied in Freysa) want Deckard dead – K's decision nonetheless gives the story a conservative-humanist twist: it tries to exempt the domain of family from the key social conflict, presenting both sides as equally brutal. This not taking sides betrays the falsity of the film: it is all too humanist, in the sense that everything revolves around humans and those who want to be (taken as) humans or those who don't know they are not humans. (Is the result of biogenetics not that we – 'ordinary' humans – effectively are that, humans who don't know they are not humans, i.e. neuronal machines with self-awareness?) The film's implicit humanist message is that of liberal tolerance: we should give androids with human feelings (love and so on) human rights, treat them like humans, incorporate them into our universe – but with their arrival, will our universe still be ours, will it remain the same human universe?

What is missing is any consideration of the change that the arrival of androids with awareness will mean for the status of humans themselves: we humans will no longer be humans in the usual sense, something new will emerge, and how are we to define it? Furthermore, with regard to the distinction between androids with 'real' bodies and hologram androids, how far should our recognition extend? Should hologram replicants with emotions and awareness (like Joi, who was created to serve and satisfy K) also be recognized as entities which act as humans? We should bear in mind the fact that Joi, ontologically a mere hologram replicant with no actual body of her own, commits the radical act of sacrificing herself for K, an act for which she was not programmed.[29]

Avoiding this New leaves the sole option of a nostalgic feeling of threat (the threatened 'private' sphere of sexual reproduction), a falsity which is inscribed into the visual and narrative form of the film. In this form, the repressed aspect of its content returns, not in the sense that the form is more progressive, but in the sense that the form serves to obfuscate the progressive anti-capitalist potential of the story. The slow rhythm, with its aestheticized imagery, expresses the social stance of not taking sides, of passive drifting.

This brings us back to the class struggle. Replicants are mass-produced slave labour for privileged humans, especially apt to work in poisonous territories on other planets where humans cannot survive – or, as Wallace says: 'Every great civilization was built off the back of a disposable workforce.' While this is true, it would nonetheless have been all too easy to reduce the antagonism between humans and replicants to a metaphorical displacement of that between the privileged and the underprivileged (exploited/excluded) within human society:

> As K goes about his investigations, we see more uncomfortable hints of a slave society supposedly taking place thirty years hence: silos of child labourers dismantling discarded electronic circuitry; scavengers living on giant scrapyards of rusting metal; female sex workers on the streets; a 'protein farmer' eking out a miserable existence in the mud. We even see a cleaner at one point – possibly the first one ever to be spotted in a Hollywood sci-fi movie. This is our world, happening now.[30]

Again, while this is true, it is not the whole truth: the prospect of posthuman forms of life is no longer just a project that concerns next generations, but something that mobilizes the ongoing attempts of global capitalism to postpone its final crisis. This brings us to what can claim to be the message of the film:

> Beyond the Baudrillardian philosophising about 'how do we know we're human?', *Blade Runner 2049* asks what it means to be human, and it boldly ventures some suggestions. It is the ability to form connections, to empathise with others, to love, to have values. It is also the will to act, to resist, to fight for those values. 'Dying for the right cause is the most human thing you can do,' says one character. It's a call to revolution. Not tomorrow but now.[31]

Here, however, ambiguity enters. The character who pronounces the statement about dying for the right cause is Freysa, the leader of the replicant resistance, and while K repeats it many times, he gives this statement a different non-political spin, replacing the emancipatory cause of universal liberation with the goal of bringing about a father–daughter reunion by extracting both of them from the political struggle – in K's reading, dying for the right cause is precisely not a call to revolution.[32]

So what would have been an authentic form of contact between a human and a replicant? When the question, 'Are androids to be treated like humans?' is debated, the focus is usually on awareness or consciousness: do they have an inner life? (Even if their memories are programmed and implanted, they can still be experienced as authentic.) Perhaps, however, we should shift the focus from consciousness or awareness to the unconscious: do they have an unconscious in the precise Freudian sense? The unconscious is not some deeper irrational dimension but what Lacan would have called a virtual 'other scene' which accompanies the subject's conscious content.

Let's take a perhaps unexpected example. Recall the famous joke from Lubitsch's *Ninotchka*: "Waiter! A cup of coffee without cream, please!' 'I'm sorry, sir, we have no cream, only milk, so can it be a coffee without milk?' At the factual level, the coffee remains the same coffee, but what we can change is to make the coffee without cream into a coffee without milk – or, even more simply, to add the implied negation and call the black coffee a coffee without milk. The difference between 'black coffee' and 'coffee without milk' is purely virtual, there is no difference in the real cup of coffee, and exactly the same goes for the Freudian unconscious: its status is also purely virtual, it is not a 'deeper' psychic reality – in short, the unconscious is like 'milk' in 'coffee without milk'. And therein resides the catch: can the digital big Other, which knows us better than we know ourselves, also discern the difference between 'black coffee' and 'coffee without milk'? Or is the counterfactual sphere outside the scope of the digital big Other, which is restricted to facts in our brain and social environs of which we are unaware? The difference we are dealing with here is that between the 'unconscious' (neuronal, social) facts that determine us and the Freudian 'unconscious', whose status is purely counterfactual. This domain of counterfactuals can only be operative if subjectivity is at work: in order to register the difference between 'black coffee' and 'coffee without milk', a subject has to be operative. And – back to *Blade Runner 2049* – can replicants register this difference?

2

Vagaries of Power

Lenin Navigating in Uncharted Territories

In his *State and Revolution* (1917), a kind of preparatory theoretical work for the October Revolution, Lenin outlined his vision of the workers' state, where every *kukharka* (not a cook, especially not a great *chef*, but more a modest woman-servant in the kitchen of a wealthy family) will have to learn how to rule the state, where everyone, even the highest administrators, will be paid the same workers' wages, where all administrators will be directly elected by their local constituencies, which will have the right to recall them at any moment, where there will be no standing army.

How this vision turned into its opposite immediately after the October Revolution is the stuff of numerous critical analyses; but what is perhaps much more interesting is the fact that Lenin proposes as the normative ground of this 'utopian' vision an almost Habermasian notion of 'the elementary rules of social intercourse that have been known for centuries and repeated for thousands of years in all copy-book maxims'[1]. In Communism, this permanent normative basis of human intercourse will finally rule in a non-distorted way: only in a Communist society, 'freed from capitalist slavery, from the untold horrors, savagery, absurdities and infamies of capitalist exploitation', will people 'gradually become accustomed to observing the elementary rules of social intercourse ... They will become accustomed to observing them without force, without coercion, without subordination, without the special apparatus for coercion called the state.'[2]

A page or so later, Lenin again writes: 'we know that the fundamental social cause of excesses, which consist in the violation of the rules of social intercourse, is the exploitation of the people';[3] does this mean that revolution is normatively grounded in some kind of universal rules that function as eternal 'human nature'? (And perhaps we find an echo of Lenin's preoccupation with 'elementary rules of social intercourse' in his critical remarks about Stalin's boorish manners made in the last months of his life.) We should pursue this question of normative grounding into all its dimensions – for example, on what are we to base our rejection of racism?

Lacan claims that science is born 'from the moment when Galileo established minute relations from letter to letter with a bar in the interval . . . this is where science takes its starting point. And this is why I have hope in the fact that, passing beneath any representation, we may perhaps arrive at some more satisfactory data on life.'[4] Jean-Claude Milner makes it clear how, for Lacan, 'relations from letter to letter, rather than mathematics, are the real point of departure': 'After a long period in which mathematics had annexed the letters in science, letters as such have now reappeared in their full autonomy. For that reason, it is possible to hope for some better data about life. Why? Because the re-emergence of autonomous letters in modern science happened in biology.'[5]

The view of existence thus posited radically precludes all the main features of our intuitive notion of life as an organic unity: 'This chemical construction which, starting from elements distributed in whatever medium and in whatever way we wish to qualify it, would build, by the laws of science alone, a molecule of DNA – how could it set off? All that science leads to is but the perception that there is nothing more real than that; in other words, nothing more impossible to imagine.'[6]

Milner draws from this irrepresentability of life, when we conceive its structure in the guise of formulas composed of letters, a radical political conclusion: only the reduction of life to a set of letters deprived of any deeper meaning or organic unity can protect us from racism:

> For many centuries, life had been the mother of all imaginary representations, the most tragic example of which had been given by the politics of race and *Lebensraum*. Thanks to the letter, it is possible to

> hope to move beyond the representations, even on the subject of life . . . If literalized, life is *the* Real as such; if biogenetics, rather than mathematics, is *the* science of the Real, then all forms of pseudo-representation that pretend to be based on life's reality lead to the fundamental myth of modern humanity, namely racism. Conversely the ultimate weapon against racism is not pity or fear, but the irrepresentability of life's lettering.[7]

Are things as simple and clear as that, however? Even James Watson, who won the Nobel prize in 1962 as one of the discoverers of the double-helix structure of DNA in 1953, repeatedly claimed that black people are less intelligent than white people, and that the idea that 'equal powers of reason' were shared across racial groups was a delusion.[8] Such claims were not just his private opinions: he based them on his work on DNA. And he was not alone in this – many racists try to ground the hierarchy of race in biogenetics. The main Rightist politician in Slovenia claims that Slovenes are genetically closer to Scandinavians than to other Slavs (his point is, predictably, to detach Slovenes from the Balkans and make them part of the northern German ethnic group). The very unrepresentability of life's lettering (mentioned by Milner) confers on racism an aura of scientific magic.

Watson's argument is not as easy to refute as it may appear. If we look at the problem from a purely scientific standpoint, why should 'powers of reason' (in whichever problematic way we define them) be equal among races? Equality is an ethico-political norm, not a fact: people are equal in spite of their natural and social differences. One should even take a step further and ask: what is the exact status of equality? What do we mean when we claim that people are equal, that they all share the same freedom, reason and dignity? If this equality as a norm is a historical fact, something that only emerged with modernity, then people only became equal when equality became a norm. So on what do we base our demand for equality? Is it a natural fact (in what sense?), a fact (or, rather, an *a priori* feature) of human nature, or (as Habermas tried to demonstrate) a normative structure implied by the fact of symbolic communication; or, again, a norm that emerges with modernity (and which, consequently, has no meaning in premodern civilizations, so that it is effectively a form of cultural colonialism to treat it as universal)? Furthermore, if the so-called axiom of equality is part of a specific historical constellation, in what sense can we claim that it is ethically superior to more traditional (or modern scientific) forms of hierarchy? Is not every claim for equality's superiority circular, in the sense of already presupposing what it tries to demonstrate? A Hegelian answer would have been that equality-in-freedom arises inevitably out of the pragmatic contradictions inherent in all previous notions of justice; but are we still ready to endorse the 'Eurocentric' notion of progress that underlies this approach?

The reference to human nature is not Lenin's last word, however. In another passage in *The State and Revolution* he seems to claim almost the opposite: surprisingly, he grounds the (in)famous difference between the lower and higher state of Communism in a different state of human nature. In the first, lower, stage we are still dealing with the same 'human nature' as in the entire history of exploitation and class struggle, while what will happen in the second, higher, stage is that 'human nature' itself will be changed:

> We are not utopians, we do not indulge in 'dreams' of dispensing *at once* with all administration, with all subordination; these anarchist dreams ... serve only to postpone the Socialist revolution *until human nature has changed*. No, we want the Socialist revolution *with human nature as it is now*, with human nature that cannot dispense with subordination, control and 'managers' ... The united workers themselves ... will hire their own technicians, managers and book-keepers, and pay them *all*, as, indeed, *every* state official, ordinary workmen's wages.[9]

The interesting point here is that the passage from the lower to the higher stage of Communism does not primarily rely on the development of productive forces beyond scarcity, but on a change in human nature. In this sense Chinese Communists (at their most radical moment) were right: there can be a Communism of poverty if we change human nature, and a socialism of relative prosperity ('goulash Communism'). When the situation is most desperate (as it was in Russia during the Civil War of 1918–20), there is always the millenarian temptation to see in this utter misery a unique chance to pass directly to Communism; Platonov's *Chevengur* has to be read against this background. Lenin thus seems to oscillate between a Habermasian reference to eternal natural rules of social exchange and a change in human nature itself, the emergence of a New Man; in what are Lenin's oscillations and tensions grounded? Let us turn to Milner's perspicuous analysis of the imbroglios of modern European revolutions which culminated in Stalinism. Milner's starting point is the radical gap that separates exactitude (factual truth, accuracy about facts) and truth (the cause to which we are committed):

> When one admits the radical difference between exactitude and truth, only one ethical maxim remains: never oppose the two. Never make of the inexact the privileged means of the effects of truth. Never transform these effects into by-products of the lie. Never make the Real into an instrument of the conquest of reality. And I would allow myself to add: never make revolution into the lever of an absolute power.[10]

In justifying this claim to absolute power, the role of proverbs is significant in the Communist tradition, from Mao's 'revolution is not a dinner party' to the legendary Stalinist 'you cannot make an omelette without breaking eggs'. The preferred saying among the Yugoslav Communists was more obscene: 'you cannot sleep with a girl without leaving some traces'. But the point made is always the same: the endorsement of brutality with no constraints. For those for whom God (in the guise of the big Other of history, whose instruments they are) exists, everything is permitted. However, theological reference can also function in the opposite way: not in the fundamentalist sense of legitimizing political measures as the imposition of divine will, whose instruments are the revolutionaries, but in the sense that the theological dimension serves as a kind of safety valve: a sign of the openness and uncertainty of the situation, which prevents the political agents from conceiving of their actions in terms of self-transparency – 'God' means we should always bear in mind that the outcome of our actions will never fit our expectations. This 'mind the gap' does not only refer to the complexity of the situation in which we intervene; above all it concerns the utter ambiguity of the exercising of our own will.

Was this short-circuit between truth and exactitude not Stalin's basic axiom (which, of course, had to remain unspoken)? Truth is not only allowed to ignore exactitude; it is allowed to refashion it arbitrarily. Perhaps the peculiarity of some Russian words can guide us in this matter: often there are two words for what appears to us Westerners the same term, one designating the ordinary meaning of the term and the other a more ethically charged 'absolute' use. There is *istina*, the common notion of truth as adequacy to facts, and (usually capitalized) *Pravda*, the absolute truth designating also the ethically committed ideal Order of the Good. There is *svoboda*, the ordinary freedom to do what we want within the existing social order, and *volja*, the more metaphysically charged absolute drive to follow one's will to the point of self-destruction – as the Russians like to say, in the West you have *svoboda*, but we have *volja*. Then there is *gosudarstvo*, the state in its ordinary administrative aspects, and *derzhava*, the state as the unique agency of absolute power. (Applying the well-known Walter Benjamin–Carl Schmitt distinction, one may venture the claim that the difference between *gosudarstvo* and *derzhava* is that between constituted and constituting power: *gosudarstvo* is the state administrative machine running its course as prescribed by legal regulations, while *derzhava* is the agent of unconditional power.) There are intellectuals, educated people, and *intelligentsia*, intellectuals charged with and dedicated to a special mission to reform society. (Along the same lines, there is already in Marx the implicit distinction between 'working class' – a simple category of social being – and 'proletariat' – a category of truth, the revolutionary subject proper.)

Is this opposition ultimately not the one, elaborated by Alain Badiou, between Event and the positivity of mere Being? *Istina* is the mere factual truth (correspondence, adequacy), while *pravda* designates the self-relating Event of truth; *svoboda* is ordinary freedom of choice, while *volja* is the resolute Event of freedom. In Russian this gap is clearly visible, and thus highlights the radical risk involved in every Truth-Event: there is no ontological guarantee that *pravda* will succeed in asserting itself at the factual level (covered by *istina*). And again, it seems as if the awareness of this gap is embedded in the Russian language, in the unique expression *awos* or *na awos*, which means something like 'with luck'; it articulates the hope that things will turn out OK when one makes a risky radical gesture without being able to foresee all its possible consequences, something like Napoleon's 'On attaque, et puis on verra,' often quoted by Lenin.

So where does Lenin stand on this? Milner locates him on the edge, bringing the tension to its extreme: while he remained fully pledged to Marxist orthodoxy, which views revolution as part of the global historical reality, in his political practice he exercised the utmost openness and improvisation, passing from revolutionary terror to a partial opening-up to capitalism; and in this process the Bolsheviks 'committed all possible mistakes', as he himself put it:

> During the French Revolution itself, it is easy to recognize the moments in which the most rational and the most courageous among the revolutionaries despaired. Most of them were competent and cultured, but no historical precedent in history, no scientific discovery, and no philosophical argument could help them. The same can be said about Lenin. Whoever has read his works cannot but admire his intelligence, his encyclopedic culture and his ability to invent new political concepts. Nonetheless, his own writings show a growing uncertainty about the situation that he himself had created. Right or wrong, the NEP was not only a turning point; it implied a severe self-criticism, bordering on renegading. At least, it proved that Lenin had been confronted by his own lack of knowledge in the field of political economy, where, as a Marxist, he was the most sure of himself; he was indeed discovering a new political country. He was encountering the very difficulty that Saint-Just had announced.[11]

This brings us back to Hegel's ambiguity with regard to political engagement: how are we to combine Hegel's after-the-fact stance (thinking is like the flight of the owl of Minerva: its aim is not to discern what is to come but to grasp the rational structure of what is) with his passionate engagement in political matters (his last writing was a polemic against the British Reform Bill of 1831)? From where, from which standpoint, did he engage? How did he avoid falling back into the Ought (*Sollen*) here, in his engaged writings? Does he not, in his critique of the Reform Bill, give way to his fear that the universal vote (in which direction the Reform Bill was moving) threatened his own model of the corporate state? So did he nonetheless judge events from the standpoint of the model of the rational state deployed in his *Philosophy of Right* – and thereby violate his own insight, according to which the very fact that he developed the model means that its time has passed? It is too easy to say that, for Hegel, 'what is' is not just a stable state of things but an open historical situation full of tensions and potentials.

It is more productive to link Hegel's impasse with Saint-Just's insight: 'Ceux qui font des révolutions ressemblent au premier navigateur instruit par son audace' ('Revolutionaries are akin to a first navigator guided by his audacity alone').[12] Isn't this the implication of Hegel's confinement of the conceptual grasp to the past? As engaged subjects, we have to act with a view to the future, but for *a priori* reasons we cannot base our decisions on a rational pattern of historical progress (as Marx thought), so we have to improvise and take risks. Was this also the lesson Lenin learned from reading Hegel in 1915? The paradox is that what Lenin took from Hegel – who is usually decried as *the* philosopher of historical teleology, of inexorable and regular progress towards freedom – was the utter contingency of the historical process.

One is tempted here to compare the 'decisionist' Lenin of 1917 with the Lenin of his last years, a more pragmatic and realist Lenin desperately trying to institutionalize revolution in a much more modest way; however, what the two stances share is, I am not afraid to say, the ruthless will to grab power and then hold onto it. Lenin's determination to take power did not just reflect his obsession with it, it meant much more – his obsession (in a good sense of the term) with opening up a 'liberated territory', a space controlled by emancipatory forces outside the global capitalist system. This is why any romantic notions of permanent revolution were totally alien to Lenin: when, after the defeat of the hoped-for pan-European revolution in the early 1920s, some Bolsheviks thought it would be better to lose power than to stick to it in those conditions, Lenin was horrified. Lenin was a kind of structuralist: the place of power has priority over its content, so we should hold onto it and then work out how to fill it in.

Furthermore, there is a clear contrast between Lenin's strategy of risking big actions and his ruthless pragmatism – a pragmatism that is evident in his decision to enforce the October Revolution. After the February Revolution, Lenin immediately saw a unique chance to take power. His insight resulted from his analysis of a very specific constellation – it was not an expression of some abstract 'decisionism'. On the other hand, there was much more 'utopianism' in Lenin's efforts to fill the free space outside the capitalist system with new content. The paradox is that he was a pragmatist in terms of how to grab power, and a utopian in terms of what to do with it.

What Lenin really learned from Hegel was the concept of concrete universality and its use in politics. 'Concrete universality' means that there is no abstract universality of rules, there are no 'typical' situations, all we are dealing with is exceptions; however, a concrete totality is the totality that regulates the concrete context of exceptions. We should thus, on account of our very fidelity to concrete analysis, reject any form of nominalism. To the nominalist claim that there is no pure neutral universality, that every universality is caught up in the conflict of particular ways of life, one should reply: 'No, today it's the particular ways of life that do not exist as autonomous modes of historical existence, the only actual reality is that of the universal capitalist system.' This is why, in contrast to identity politics, which focuses on how each (ethnic, religious, sexual) group should be able fully to assert its particular identity, the much more difficult and radical task is to enable each group to access full universality. This access to universality does not mean a recognition that one is also part of the universal human genus, or the assertion of some ideological values that are considered universal. Rather, it means recognizing one's own universality, the way it is at work in the fractures of one's particular identity, as the 'work of the negative' that undermines every such identity.

Paradoxically, Lenin's philosophical weakness contributed to – and was even a condition of – his political genius. So although Lukács in the early 1920s (in his *History and Class Consciousness* and *Lenin*) was right to interpret Lenin's thought and action as being grounded in the structure of Hegelian subjectivity, with the proletariat as the historical subject-substance, it was not clear to him that, for complex reasons of historical dialectics, a Lenin fully aware of what he was doing would not be able to do it. Here is another case of the strange dialectic of not-knowing as a condition of doing; and the surprise is that this example occurs in the work of Lukács, a philosopher whose notion of class consciousness implies the self-transparent identity of knowing and doing (the very act of arriving at class consciousness is for the proletariat a practical act, a doing, a simultaneous change in its actual social being).[13] No wonder, then, that although Lenin tried to develop a theoretical framework for this practice (a framework of complex, overdetermined totality in which exception *is* the law and allows for a revolution in the 'weakest link' of the capitalist system), the tension became increasingly palpable.

So what did Stalin do here? 'Stalin chose the easy way in preferring the absolute solitude of S1 [the Master-Signifier] which leads to absolute opportunism. No party, no family, no allies except circumstantial ones, but also no predetermined theory of social forms, no accepted criteria for rationality, no ethical rules.'[14]

Perhaps Milner's reading is a little bit too narrow here. At a certain level, Stalin's break with Lenin was purely discursive, violently imposing a radically different subjective economy. The gap between general principles ('historical laws') regulating reality and pragmatic, improvised decisions still discernible in Lenin is simply disavowed, and the two extremes directly coincide: on the one hand, we get total pragmatic opportunism; on the other, this pragmatic opportunism is legitimized by a new Marxist orthodoxy that proposes a general ontology. What this means is that Lenin himself was not a 'Leninist': 'Leninism' is a retroactive construction of Stalinist discourse. The key to Leninism as (Stalinist) ideology is provided by Mikhail Suslov, the member of the Politburo responsible for ideology from Stalin's later years up to the Gorbachev era. Alexei Yurchak pointed out that neither Khrushchev nor Brezhnev would release any document until Suslov had looked over it – why?

> In 1990, Fyodor Burlatsky, a former advisor to Khrushchev and Andropov, described a technique that Suslov used to manipulate Lenin's words. Suslov, who occupied the position of the Politburo's

> head of ideology, had an enormous library of Lenin's quotes in his Kremlin office. They were written on library cards, organized by themes, and contained in wooden file cabinets. Every time a new political campaign, economic measure, or international policy was introduced, Suslov found an appropriate quote from Lenin to support it. Once in the early 1960s, young Burlatsky showed Suslov a draft of a speech he prepared for Khrushchev. Having carefully studied the text, Suslov pointed to one place and said: 'It would be good to illustrate this idea with a quote from Vladimir Il'ich.' When Burlatsky replied that he would find an appropriate quote, Suslov interrupted: 'No, I will do this myself.' Burlatsky writes: 'Suslov dashed to the corner of his office, pulled out one drawer and put it on the table. With his long, thin fingers he started very rapidly flipping through the cards. He pulled out one and read it. No, that's not it. Then he pulled out another one. No, still not right. Finally he took another card out and exclaimed with satisfaction, 'Ok, this one will do.'
>
> Lenin's quotes in Suslov's collection were isolated from their original contexts. Because Lenin was an extremely prolific writer who commented on all sorts of historical situations and political developments, Suslov could find appropriate quotes to legitimate as 'Leninist' almost any argument and initiative, sometimes even if they opposed each other. Another writer remembered that 'the very same quotes from the founders of Marxism-Leninism that Suslov successfully used under Stalin and for which Stalin so highly valued him, Suslov later employed to critique Stalin'.[15]

This was the truth of Soviet Leninism. Lenin served as the ultimate reference: a quote of his legitimized any political, economic or cultural measure, but in a totally pragmatic and arbitrary way – in exactly the same way, incidentally, that the Catholic Church referred to the Bible. (One should also consider to what extent Lenin himself used Marx's texts in a similar way.) In other words, the reference to Lenin posed no boundaries whatsoever: any political measure was acceptable if legitimized by a quote of his. Marxism thus becomes a 'world-view' allowing us access to objective reality and its laws, and this process brings a new false sense of security: our acts are 'ontologically' covered, part of 'objective reality' regulated by laws known to us

Communists. However, the price paid for this ontological security is terrible: exactitude, in the sense of truth about facts, to which Lenin was still committed, disappears – facts can be voluntarily manipulated and retroactively changed, events and persons become non-events and non-persons. In other words, in Stalinism the Real of politics – brutal, subjective interventions that violate the texture of reality – returns with a vengeance, although in the form of its opposite, respect for objective knowledge.

Following the Stalinist turn, Communist revolutions were grounded in a clear vision of historical reality ('scientific socialism'), its laws and tendencies, so that, in spite of the unpredictable turn of events, the revolution was fully located in this process of historical reality – as they liked to say, socialism should be built in each country according to its particular conditions, but in accordance with general laws of history. In theory, revolution was thus deprived of the dimension of subjectivity proper, of radical incursions of the Real into the texture of 'objective reality' – in clear contrast to the French Revolution, whose most radical figures perceived it as an open process lacking any support in a higher necessity.

Today, even more than in Lenin's time, we navigate in uncharted territories, with no global cognitive mapping – but what if it is this very lack of clear mapping that gives us hope for a way to avoid totalitarian closure?[16]

Elections, Popular Pressure, Inertia

Yanis Varoufakis opens his *Adults in the Room* with a report on how, on 16 April 2015, in a dark corner of a DC hotel bar, Larry Summers told him:

> 'Yanis, you made a big mistake.'
> Faking steeliness, I replied, 'And what mistake was that, Larry?'
> 'You won the election!' came his answer.[17]

In what precise sense was the electoral victory of Syriza a

mistake? In accepting the electoral game, in winning at the wrong moment or . . .? The second round of the French presidential elections in May 2017 confronted us even more strongly with this old dilemma of the radical Left: vote or not in the parliamentary elections? The miserable choice of Marine Le Pen or Emmanuel Macron exposes us to the temptation of ceasing to vote altogether, of refusing to participate in this increasingly meaningless ritual.

Making a decision on this is full of ambiguities. The argumentation against voting subtly (or openly) oscillates between two versions, the 'soft' one and the 'strong' one. The 'soft' version specifically targets the multi-party democracy in capitalist countries with two main arguments: (1) the media controlled by the ruling class manipulate the majority of voters and do not allow them to make rational decisions in their own interest; (2) elections are a ritual that occurs every four years and their main function is to passivize voters in the long periods between elections. The ideal that underlies this critique is that of a non-representative 'direct' democracy, with continuous direct participation of the majority. The 'strong' version takes a crucial further step and relies (explicitly or not) on a profound distrust of the majority of people: the long history of universal suffrage in the West shows that the vast majority of people are as a rule passive, caught up in the inertia of survival, not ready to be mobilized for a cause. That's why every radical movement is always constrained to a vanguard minority, and in order to gain hegemony it has to wait patiently for a crisis (usually war) which provides a narrow window of opportunity. In such moments, an authentic vanguard can seize the day, mobilize the people (even if not the actual majority) and take over.

Communists were always utterly 'non-dogmatic', ready to be parasitic on other issues: land and peace (Russia), national liberation and unity against corruption (China). They were always well aware that mobilization would soon be over, and were carefully preparing the apparatus to keep themselves in power at that moment. (In contrast to the October Revolution, which explicitly treated peasants as secondary allies, the Chinese Revolution didn't even pretend to be proletarian: it directly addressed farmers as its core of support.)

The big, defining problem of Western Marxism was the insufficiently motivated revolutionary subject: how is it that the working class does not complete the passage from in-itself to for-itself and constitute itself as a revolutionary agent? This problem provided the main *raison d'être* for Marxism's reference to psychoanalysis, which was invoked to explain the unconscious libidinal mechanisms that prevent the rise of class consciousness that is intrinsic to the very being, the social situation, of the working class. In this way, the truth of Marxist socio-economic analysis was preserved, and there was no reason to cede ground to 'revisionist' theories about the rise of the middle classes, etc. For this same reason, Western Marxism was also constantly on the lookout for other social agents who could play the revolutionary role, the understudy replacing the indisposed working class: Third World peasants, students and intellectuals, the excluded ... and now the refugees.

The failure of the working class as a revolutionary subject was at the very heart of the Bolshevik Revolution: Lenin's skill was to detect the 'rage potential' of the disaffected peasants. The October Revolution was won with the slogan 'land and peace', addressed to the vast peasant majority, seizing the brief moment of their radical dissatisfaction. Lenin had already been thinking along these lines a decade previously, which is why he was horrified at the prospect of the success of the Stolypin land reforms, aimed at creating a new, strong class of independent farmers – he wrote that if Stolypin succeeded, the chance for a revolution would be lost for decades.

All successful socialist revolutions, from Cuba to Yugoslavia, followed this model, grasping the opportunity in an extreme critical situation, co-opting the national-liberation struggle or other movements fed by 'rage capital'. Of course, a believer in the logic of hegemony would here point out that this is the 'normal' logic of revolution, that the 'critical mass' is reached precisely and only through a series of equivalences among multiple demands, which is always dependent on a specific, unique even, set of circumstances. A revolution never occurs when all antagonisms collapse into the big One, but when they synergetically combine their power.

The problem is here more complex, however: the point is not just that revolution no longer rides the train of history, following its laws, since there is no history, history being a contingent, open process. The problem is a different one: it is as if there *is* a Law of History, a more or less clear, predominant main line of historical development, and that revolution can only occur in its interstices, 'against the current'. Revolutionaries have to wait patiently for the (usually very brief) period of time when the system openly malfunctions or collapses, then seize the window of opportunity, grasp the power that at that moment lies on the street, as it were, and is up for grabs, and then tighten their hold on that power, building repressive apparatuses and so on. Then, once the moment of confusion is over, and the majority have sobered up and are disappointed by the new regime, it is too late to get rid of it and it is firmly entrenched.

Not only this, but Communists have always been careful to calculate the right moment to stop popular mobilization. Let's take the case of the Chinese Cultural Revolution, which undoubtedly contained elements of an enacted utopia. At its very end, before the agitation was blocked by Mao himself (because he had already achieved his goal of re-establishing full power and getting rid of the top *nomenklatura* competition), there was the 'Shanghai Commune', discussed in the Introduction. It is significant that it was at this very point that Mao ordered the army to intervene and restore order. The paradox is that of a leader who triggers an uncontrolled upheaval while trying to exert full personal power – the overlapping of extreme dictatorship and extreme emancipation of the masses.

The most visible aspect of 'popular presence' is thus the assemblage of large groups of people in central public spaces, and an important open question is: how does cyberspace presence/pressure operate, what is its potential? Popular presence is precisely what the term says – presence as opposed to representation, pressure directed at representative organs of power; it is what defines populism in all its guises, and (as a rule, although not always) it has to rely on a charismatic leader. Examples abound: the crowd outside the Louisiana congress that supported the populist governor Huey Long and assured his victory in a key vote in 1932, crowds exerting pressure on behalf of Milošević in Serbia, crowds persisting for days in Tahrir

Square during the Arab Spring demanding the overthrow of Mubarak, crowds in Istanbul during protests against Erdoğan, and so on. In a popular presence, 'people themselves' make palpable their force directly and beyond representation, but at the same time they become another mode of being. In a short poem written about the GDR workers' uprising in 1953, Brecht quotes a contemporary Party functionary as saying that the people have lost the trust of the government. Would it not therefore be easier, Brecht slyly asks, to dissolve the people and have the government elect another one? Instead of reading this poem as ironic, one should take it seriously: yes, in a situation of popular mobilization, the inert mass of ordinary people is transubstantiated into a politically engaged united force.

One should always bear in mind that a permanent people's presence means a permanent state of emergency – so what happens when people get tired, when they are no longer able to sustain the tension? Communists in power had two solutions (or, rather, two sides of one and the same solution): the Party's reign over a passive population and fake popular mobilization. Trotsky himself, the theorist of the permanent revolution, was well aware that people 'cannot live for years in an uninterrupted state of high tension and intense activity',[18] and he turns this fact into an argument for the need for the vanguard Party: self-organization in local councils cannot take over the role of the Party, which should run things when the people get tired; and, to amuse the people and to maintain appearances, an occasional big spectacle of pseudo-mobilization can be of some use, from Stalinist parades to today's in North Korea. In capitalist countries there is, of course, another way to ease popular pressure: (more or less) free elections – recently in Egypt and Turkey, but they also worked in 1968 in France. One should never forget that the agent of popular pressure is always a minority – the number of active participants in the Occupy Wall Street movement of 2011 against global economic equality was much closer to 1 per cent than to the 99 per cent of its slogan.

The French language uses the so-called *ne explétif* after certain verbs and conjunctions. It is also called the 'non-negative *ne*' because it has no negative value in and of itself – it is used in situations where the main clause has a negative (either negative-bad or negative-negated) meaning, such as expressions of fear, warning, doubt and negation.[19] For example: 'Elle a peur qu'il ne soit malade' (She's afraid that he is sick). Lacan noted how this superfluous negation renders perfectly the gap that separates our true unconscious desire from our conscious wish: when a wife is afraid that her husband is sick, she may well worry that he is not sick (desiring him to be sick). And could we not say exactly the same about the perennial complaint of ruling parties in socialist countries that ordinary people are not engaged enough in political activity, that they are too passive and indifferent? 'Ils ont peur que le people ne soit passif et indifférent' – what they really fear is that ordinary people will *not* remain passive and indifferent.

Should we just ignore elections, then? Whatever elections are, they measure something in a purely numeric way – the percentage of the population that stands behind the main publicly presented political options. That's why Communists in power always have to stick to the form of free secret elections, even if the outcome is a totally predictable 90 per cent or more in favour of the existing regime (after two years in power, even the Khmer Rouge performed this ritual); or, even more so, to the form of multi-party democracy, as formerly in Poland and the GDR – how many people are aware that even China is today a multi-party democracy with seats allotted to other 'patriotic' forces apart from the Communist Party? Plus, isn't some kind of election necessary to form the leading body of the ruling party itself? This was the great problem with early Bolshevism: is it possible to have inner-Party democracy without some kind of democracy in the society outside the Party? How do you keep open the space for authentic feedback from the people outside the Party circle? The problem was never that the Party *nomenklatura* didn't know what the people really thought – through their secret services they were always all too well informed about it.

The Chinese model is the most consistent one in this regard: members of the *de facto* ruling body (seven members of the Standing Committee of the Political Bureau of the Communist Party of China) are elected at a Party congress every eight years or so, and there is no debate – at the end of the congress they are simply presented as a mysterious revelation. The selection procedure involves complex and totally opaque behind-the-scenes negotiations, so that the assembled delegates who unanimously approve the list learn about it only when they vote. We are not dealing here with some kind of secondary 'democratic deficit': this impenetrability is structurally necessary (within the authoritarian system, the only alternatives are a *de facto* monarchy, as in North Korea, or the traditional Communist model of a leader who only stops ruling when he dies).

The basic problem is this: how to move beyond multi-party democracy without falling into the trap of direct democracy? In other words: how to invent a different mode of *passivity* of the majority, how to cope with the unavoidable *alienation* of political life? This alienation has to be understood at its most basic level, as the excess constitutive of the functioning of an actual power, overlooked by liberalism as well as by Leftist proponents of direct democracy. Recall the traditional liberal notion of representative power: citizens transfer part of their power to the state, but on precise terms (this power is constrained by law, limited to very precise conditions in the way it is exercised, since the people remain the ultimate source of sovereignty and can withdraw power if they decide so). In short, the state with its power is the minor partner in a contract that the major partner (the people) can at any point annul or change, basically in the same way each of us can change the supermarket where we do our shopping. However, the moment we take a closer look at an edifice of state power, we can easily detect an implicit but unmistakable message: 'Forget about our limitations – ultimately, we can do whatever we want with you!' This excess is not an adjunct polluting the purity of power, but a necessary constituent – without it, without the threat of arbitrary omnipotence, state power is not a true power, it loses its authority.

The way to break the spell of power is thus not to succumb to the fantasy of a transparent power; one should rather hollow out the edifice of power from within by separating the edifice from its agent (the bearer of power). As developed decades ago by Claude Lefort, therein resides the core of 'democratic invention', in the empty place of power, the gap between the place of power and the contingent agents who, for a limited period, can occupy that place. Paradoxically, the underlying premise of democracy is thus not only that there is no political agent that has a 'natural' right to power, but, much more radically, that 'the people' itself, the ultimate source of the sovereign power in democracy, doesn't exist as a substantial entity. In the Kantian sense, the democratic notion of 'people' is a negative one, a concept whose function is merely to designate a certain limit: it prohibits any determinate agent from ruling with full sovereignty. (The only moment when 'the people' exists is at democratic elections, which is precisely that of the disintegration of the entire social edifice – in elections, 'people' are reduced to a mechanical collection of individuals.) The claim that the people *does* exist is the basic axiom of 'totalitarianism', and the mistake of 'totalitarianism' is strictly homologous to the Kantian misuse ('paralogism') of political reason: 'the people exists' through a determinate political agent which acts as if it embodies (not only re-presents) the people, its true will (the totalitarian party and its leader) – i.e. in terms of a transcendental critique, as a phenomenal embodiment of the noumenal people.

Critics of representative democracy endlessly vary the motif of how, for *a priori* formal reasons and not just on account of accidental distortions, multi-party elections betray true democracy – but, while accepting this critical point as the price to be paid for any functioning democracy, one should add that it is *because of* such minimal 'alienation' signalled by the term 'representative' that a democracy functions. That is to say, what this 'alienation' points towards is the 'performative' character of democratic choice: people do not vote for what they want (they know that in advance) – it is through such choice that they discover what they want. A true leader does not just follow the wishes of the majority; she or he makes the people aware of what they want.

This is why democracy retains its meaning even if the choice it gives is between very similar programmes – such empty choice makes it clear that there is no predestined bearer of power. The logical implication of this premise is Kojin Karatani's idea of combining elections with a lottery in determining who will rule us. This idea is more traditional than it may appear (Karatani himself mentions ancient Greece) – paradoxically, it fulfils the same task as Hegel's theory of monarchy. Karatani takes a heroic risk in proposing a crazy-sounding definition of the difference between the dictatorship of the bourgeoisie and the dictatorship of the proletariat: 'If universal suffrage by secret ballot, namely, parliamentary democracy, is the dictatorship of the bourgeoisie, the introduction of lottery should be deemed the dictatorship of the proletariat.'[20] Wasn't this also Lenin's underlying idea when, in his *The State and Revolution*, he outlined his vision, mentioned earlier, of a workers' state where every *kukharka* would have to learn how to rule the state? From (electoral) democracy to lotocracy . . .

Does this mean that expertise doesn't matter? No, since another separation enters the frame here: that between S1 and S2, between Master-Signifier and expert knowledge. The Master (people through voting) decides, makes the choice, but the experts suggest what to choose – people want the appearance of choice, not real choice-making. This is how our democracies function – with our consent: we act *as if* we are free and freely deciding, silently not only accepting but even *demanding* that an invisible injunction (embedded in the very form of our free speech) tells us what to do and think. As Marx knew long ago, the secret is in the form itself. In this sense, in a democracy, every ordinary citizen is effectively a king – but a king in a constitutional democracy, a king who only formally decides, whose function is to sign measures proposed by an executive administration. This is why the problem of democratic rituals is homologous to that of constitutional democracy: how do we protect the dignity of the king? How do we maintain the appearance that the king effectively decides, when we all know this is not true? What we call a 'crisis of democracy' does not occur when people stop believing in their own power but, on the contrary, when they stop trusting the elites, those who are supposed to know for them and provide their guidelines, when they experience the anxiety signalling that 'the (true) throne is empty', that the decision is now *really* theirs. There is thus in 'free elections' always a minimal aspect of politeness: those in power politely pretend that they do not really hold power, and ask us to freely decide if we want to restore them to power – in a way which mirrors the logic of a gesture meant to be refused.

But how is this different from 'totalitarian' Communism, where voters are also compelled to go through the empty ritual of freely choosing – voting for – what is imposed on them? The obvious answer is that in democratic elections there is a minimal free choice, a choice which minimally matters. But a more important difference is that in 'totalitarian' Communism the gap between Master-Signifier and expert knowledge disappears – how? The distance between Lenin and Stalin concerns precisely this point. So where do we stand today with regard to this dilemma?

Welcome to the Boredom of Interesting Times!

There is an old Chinese curse (about which nobody in China knows anything, so it's probably a Western invention) which says: 'May you live in interesting times!' Interesting times are times of trouble, confusion and suffering. And it seems that in some 'democratic' countries, we of late have witnessed a weird phenomenon which proves that we do live in interesting times: a candidate emerges and wins elections as it were from nowhere, in a moment of confusion building a movement around his or her name – both Silvio Berlusconi and Macron exploded on the scene like this. But what is this process a sign of? Definitely not of any kind of direct popular engagement beyond party politics – on the contrary, we should never forget that such figures emerge with the full support of the social and economic establishment. Their function is to obfuscate actual social antagonisms – people are magically united against some demonized 'fascist' threat. In 1990, Václav Havel was the first to blurt out this dream: when, after being elected President, he first met Helmut Kohl, he made a weird suggestion: 'Why don't we work together to dissolve all political parties? Why don't we set up just one big party, the Party of Europe?' One can imagine Kohl's sceptical smile . . .

What this sudden rise of parties from nowhere and with no clear programme indicates is the disintegration of the political space as we knew it: even if political parties remain the general frame of political life, it is as if they have exhausted their potential. Multiple options are possible here, which all hinge on how the new political antagonisms – populists versus technocrats, etc. – articulate themselves. But what is clear is that there is no longer a proper political project to mobilize and organize people. If a new political project is to arise, it will have to come from the Left.

This weird phenomenon, one of the visible effects of the aforementioned long-term rearrangement of the political space in Europe, brings us back to Berlusconi and Macron: new movements emerge out of nowhere when none of the old big parties, conservative or liberal, succeeds in imposing itself as the agent of the new radical centre, so the establishment is caught in a panic and has to invent a new movement in order to keep things the way they are. The names of their respective movements (more than just parties) sound similar in their empty universality, which fits everyone and everything. Who wouldn't agree with *Forza Italia!* or with *La République En Marche* – both of which slogans designate the abstract sense of a victorious movement without any specification of the direction of the movement and its goal. Both came to the fore as reactions of the establishment in a panic. There is, of course, an obvious difference between the two, a different accent: Berlusconi entered the scene when, after the big anti-corruption campaign, the entire traditional political configuration in Italy collapsed and ex-Communists remained the only viable force, while Macron came to prominence when Le Pen's neo-fascist populism threatened to win the elections. His role is best described by a word used by some of his supporters: in recent years, Marine Le Pen has gradually been de-diabolized, perceived as part of the 'normal' (acceptable) political space, and the task is to *re-diabolize* her, to show the political public that she remains the same old anti-Semitic fascist, and as such is not to be tolerated by any serious democrat. Such a gesture of re-diabolization is clearly not enough: instead of just focusing on the diabolic fetish, one should immediately ask the question of how such a 'devil' could have emerged in our society (in Le Pen's case, she is a reaction to the policy whose embodiment is Macron). The function of diabolization is precisely to obfuscate this link, to locate the guilt in an agent outside our democratic space.

A classic liberal argument for voting for Clinton or Macron against Trump or Le Pen is that while it is true that what Clinton and Macron stand for is the very predicament that gave birth to Trump or Le Pen, not voting for Clinton or Macron is like voting for an actual disaster in order to prevent a possible future disaster. This argument sounds convincing, on condition that we ignore temporality. If Le Pen had been elected President in 2017, it could have triggered strong anti-fascist mobilization, rendering her re-election unthinkable, plus it could have given a strong push to the Leftist alternative. So the two disasters (Le Pen President now or the threat of Le Pen as President in five years) are not the same: the disaster after five years of Macron's reign, if it turns out to be a failure, will be much more serious than the one which did not happen in 2017.

Historically, it was the task of the Left to ask such questions, so no wonder that, with the enemy diabolized, the radical Left conveniently disappears from the picture – recall how, in the 2017 elections in France, every Leftist scepticism about Macron was immediately denounced as support for Le Pen. So we can safely venture the hypothesis that this elimination of the Left was the true aim of the operation, and that the demonized enemy was a convenient prop. Julian Assange recently wrote that the reason why the US Democratic Party's establishment has embraced the 'We didn't lose – Russia won' narrative is because if they didn't, then the insurgency created by Bernie Sanders during the 2017 presidential election would dominate the party. And in the same way that the US Democratic establishment diabolizes Trump to get rid of Sanders, who poses a threat, the French establishment diabolized Le Pen to head off potential Leftist radicalization.

The title of Hadley Freeman's comment in the *Guardian*, the UK voice of the anti-Assange-pro-Hillary liberal Left, says it all: 'Le Pen is a far-right Holocaust revisionist. Macron isn't. Hard choice?'[21] Predictably, the text proper begins with: 'Is being an investment banker analogous with being a Holocaust revisionist? Is neo-liberalism on a par with neo-fascism?', and mockingly dismisses even the conditional Leftist support for the second-round Macron vote, the stance of: 'I'll now vote Macron – *very* reluctantly.' This is liberal blackmail at its worst: one should support Macron unconditionally, it doesn't matter that he is a neo-liberal centrist, just that he is against Le Pen . . . It's the old story of Hillary versus Trump: in the face of the fascist threat, we should all gather around her banner (and conveniently forget how her side brutally outmanoeuvred Sanders and thus contributed to losing the election). Are we not allowed at least to ask the question: yes, Macron is pro-European – but what kind of Europe does he personify? The very Europe whose failure feeds Le Pen populism, the anonymous Europe in the service of neo-liberalism! *This* is the crux of the affair: yes, Le Pen is a threat, but if we throw all our support behind Macron, do we not get caught up in a kind of circle and fight the effect by way of supporting its cause? This brings to mind a chocolate laxative available in the US. It is publicized with the paradoxical injunction: 'Do you have constipation? Eat more of this chocolate!' – in other words, eat the very thing that causes constipation in order to be cured of it. In this sense, Macron is the chocolate-laxative candidate, offering us as a cure the very thing that caused the illness.

Both candidates presented themselves as anti-system, Le Pen in an obvious populist way and Macron by a more much interesting means: he was outside the existing political parties but, precisely as such, he stood for the system as such, in its indifference to established political choices. In contrast to Le Pen, who stood for proper political passion, for the antagonism of Us against Them (from immigrants to non-patriotic financial elites), Macron represented apolitical, all-encompassing tolerance. We often hear the claim that Le Pen's politics draws its strength from fear (fear of immigrants, of the anonymous international financial institutions), but does the same not hold for Macron? He finished first because voters were afraid of Le Pen, and the circle is thus closed: neither of the candidates offered a positive vision, they were both candidates of fear.

The true stakes of this vote become clear if we locate it in its larger historical context. In Western and Eastern Europe, there are signs of a long-term rearrangement of the political space. Till recently, it was dominated by two main types of party addressing the entire electoral body – a Right-of-centre party (Christian democrat, liberal-conservative, people's) and a Left-of-centre party (socialist, social-democratic) – with smaller parties addressing a narrow part of the electorate (ecologists, neo-fascists, etc.). Now, progressively, a type of party is emerging that stands for global capitalism, usually with relative tolerance towards abortion, gay rights, religious and ethnic minorities, and so on; in opposition to this party there is typically an increasingly strong anti-immigrant populist party which, on its fringes, is accompanied by directly racist neo-fascist groups. The exemplary case is here Poland: after the disappearance of the ex-Communists, the main parties are the 'anti-ideological' centrist liberal party of the former Prime Minister Donald Tusk and the conservative Christian party of the Kaczynski brothers. The key issue for the radical centre today is this: which of the two main parties, conservatives or liberals, will succeed in presenting itself as embodying the post-ideological non-politics as against the other party, which is dismissed as being 'still caught up in old ideological spectres'? In the early 1990s, conservatives were better at it; later it was liberal Leftists who seemed to be gaining the upper hand, and Macron is the latest figure to represent a pure radical centre. Juergen Habermas and Peter Sloterdijk, the two great philosophical opponents on the German scene, have recently rendered public their admiration for Macron in quite enthusiastic terms, as a new hope for Europe, even with hints that, in the way that Hegel, when he saw Napoleon riding on a horse in Jena, described him as the *Weltgeist* riding on a horse, the same holds for Macron, who is the embodiment of today's European *Geist*. When such radical opponents starts to speak the same language, it is always a symptomatic point – not of their deeper unity but of the rejection ('repression') that unites them – the rejection of a more radical Left, in this case.

We have thus reached the lowest point in our political lives: a pseudo-choice, if ever there was one. Yes, the victory of Le Pen would pose dangerous possibilities. But the assuagement brought by Macron's triumphant victory is no less dangerous, since his victory did not really awaken us – its effect is quite the opposite: sighs of relief everywhere, the nightmare is over, thank God the danger was kept at bay, Europe and our democracy are saved, so now we can go back to our liberal-capitalist sleep again.

In the hopeless situation we are in, facing a false choice, we should summon up the courage simply to abstain from voting. Abstain, and begin to think. The commonplace 'enough talking, let's act' is deeply deceiving – now we should say precisely the opposite: enough of the pressure to do something, let's begin to talk seriously, that is, to think! And by this I mean we should also leave behind the radical Leftist self-complacency of endlessly repeating how the choices we are offered in the political space are false, and how only a renewed radical Left can save us – yes, in a way, but why, then, does this Left not emerge? What vision has the Left to offer that would be strong enough to mobilize people? We should never forget that the ultimate cause of the vicious cycle of Le Pen and Macron in which we are caught is the disappearance of a viable Leftist alternative.

No wonder a new spectre is haunting liberal-progressive politics in Europe and the US, the spectre of fascism. Trump in the US, Le Pen in France, Orbán in Hungary – they are all demonized as the new Evil against which we should unite our forces. Every minimal doubt and reservation about the alternatives we are presented with is immediately proclaimed a sign of secret collaboration with fascism. In a remarkable interview with *Der Spiegel* published in October 2017, Emmanuel Macron made some statements which were received enthusiastically by all who want to fight the new fascist Right:

> There are three possible ways to react to right-wing extremist parties. The first is to act as though they don't exist and to no longer risk taking political initiatives that could get these parties against you. That has happened many times in France and we have seen that it doesn't work. The people that you are actually hoping to support no longer see themselves reflected in your party's speeches. And it allows the right wing to build its audience. The second reaction is to chase after these right-wing extremist parties in fascination.
>
> *Der Spiegel*: And the third possibility?
>
> *Macron*: To say, these people are my true enemies and to engage them in battle. Exactly that is the story of the second round of the presidential election in France.[22]

While Macron's stance is commendable, it is crucial to supplement it with a self-critical turn. The demonized image of a fascist threat clearly serves as a new political fetish, in the simple Freudian sense of a fascinating image whose function is to obfuscate the true antagonism. Fascism itself is inherently fetishist, it needs a figure like that of a Jew, condemned as the external cause of our troubles – such a figure enables us to obfuscate the immanent antagonisms that cut across our society. My claim is that exactly the same holds for the notion of 'fascist' in today's liberal imagination: it enables us to obfuscate immanent deadlocks which lie at the root of our crisis. The desire not to make any compromises with the alt-right can easily obscure the degree to which we are already compromised by it. One should welcome every sign of this gradually emerging self-critical reflection which, while remaining thoroughly anti-fascist, also casts a critical glance at the weaknesses of the liberal Left – see, for example, the extraordinary intervention of Susan Sarandon.[23] Her claim is not that 'me too' style political correctness goes too far, but that it is pseudo-radical, that its radicality is a pose. The task is not to build a coalition between the radical Left and the fascist Right, but to cut off the working-class oxygen supply to the alt-right by addressing their voters. The way to achieve this is to move more to the Left with a more radical, critical message, i.e. to do exactly what Sanders and Corbyn were doing and which was the root of their relative success.

Another aspect of the new wave of racism is the mobilization of the obscene underside of ideology. When the black conservative Ben Carson was competing to become Republican presidential candidate, he presented his life story as the progress from a juvenile delinquent to a moral Christian. However, when journalists probed into his past, they were surprised to discover that he never was a delinquent: he had all the time been a modest, well-behaved boy. But now comes the true surprise: in response to this discovery, Carson's propagandists stressed that Carson effectively *was* a delinquent in his youth – why this weird insistence? Wouldn't it have been better for Carson to come across in the eyes of his supporters (mostly white Christian conservatives) as a good boy from the beginning? No: his delinquent past perfectly fitted his image, that of the usual black boy, caught up in crime and other vices, who found strength in hard work, discipline and Christianity. This is what his supporters wanted to see: not simply a good black boy (as such he would have to be recognized as one of us, fully equal to us), but somebody who first fully enjoyed being black in its transgressive aspects (the sins of 'lower' races always fascinate white conservatives and are clearly an object of ambiguous envy), and then found strength to castigate his black wildness and become a moral Christian like them. Recall that Carson also claimed that slavery, deplorable as it was, helped blacks to discover and accept Christianity: the role of Christianity in this story was to civilize the savage blacks by way of integrating them into white culture.

It is only against this background that we can understand how Donald Trump, a lewd and depraved person, the very opposite of Christian decency, can function as the chosen hero of the Christian conservatives. The explanation one usually hears is that, while Christian conservatives are well aware of the problematic character of Trump's personality, they have chosen to ignore this side of things since what really matters to them is Trump's agenda, especially his anti-abortion stance. If he succeeds in imposing new conservative members of the Supreme Court who will then overturn Roe vs. Wade (the Supreme Court decision that legitimized abortion), then this will wipe out all his sins . . . but are things as simple as that? What if the very duality of Trump's personality – his high moral stance, accompanied by personal lewdness and vulgarity – is what makes him attractive to Christian conservatives, what if they secretly identify with this very duality? Exactly the same goes for Poland's current *de facto* ruler, Jarosław Kaczyński, who, in a 1997 interview for *Gazeta Wyborcza*, inelegantly exclaimed, '*Teraz kurwa my*'. This phrase (which then became a classic *locus* in Polish politics) can be vaguely translated as 'It's our fucking time, now we are in power, it's our turn,' but its literal meaning is more vulgar, something like 'Now it's our time to fuck the whore' (after waiting in line in a brothel).[24] It's significant that this phrase was publicly uttered by a devout Catholic conservative, the protector of Christian morality: it's the hidden obverse that effectively sustains Catholic 'moral' politics.

The Communist side too is not far behind in similar vulgarities. For example, in his speech at the Lushan Party Conference in July 1959, when the first reports made it clear what a fiasco the Great

Leap Forward was, Mao called on the Party cadre to assume their share of the responsibility, and he concluded his speech with an admission of his own responsibility, especially for the unfortunate campaign to make steel in every village: 'The chaos caused was on a grand scale and I take responsibility. Comrades, you must all analyse your own responsibility. If you have to shit, shit! If you have to fart, fart! You will feel much better for it.'[25]

Why this vulgar metaphor? In what sense can the self-critical admission of one's responsibility for serious mistakes be compared to the need to shit and fart? I presume the answer is that, for Mao, taking responsibility does not mean an expression of remorse, which may even push you to offer to step down; it's more that, by doing it, you rid yourself of responsibility, so no wonder you 'feel much better for it', like after a good shit – you don't admit you *are* shit, you get rid of the shit in you. This is what Stalinist 'self-criticism' effectively amounts to.

The important lesson here is that this opening-up of the obscene background to our ideological space (to put it somewhat simplistically, the fact that we can now more publicly make racist, sexist, etc., statements which, till recently, belonged to our private space) in no way means that the time of mystification is over, that now ideology openly displays its cards. On the contrary, when obscenity penetrates the political scene, ideological mystification is at its strongest: the true political, economic and ideological stakes are more invisible than ever. In short, public obscenity is always sustained by a concealed moralism, its practitioners secretly believe they are fighting for a cause, and it is at this level that they should be attacked. The problem is not that Trump is a clown. The problem is that there is a programme behind his provocations, a method in his madness. Trump's (and others') vulgar obscenities are part of their populist strategy to sell this programme to ordinary people, a programme which (in the long term, at least) works against ordinary people: lower taxes for the rich, less healthcare and workers' protection, and so on. Unfortunately, people are ready to swallow many things if they are presented to them with laughter.

There is a delicious old Soviet joke about Radio Erevan (the

Armenian radio station that was the traditional butt of Soviet jokes). A listener asks: 'Is it true that Rabinovitch won a new car in the lottery?' The radio answers: 'In principle yes, it's true, only it wasn't a new car but an old bicycle, and he didn't win it, it was stolen from him.' The same goes for the French presidential elections of 2017: is it true that, in a great display of anti-fascist unity, the people of France elected an outsider and defeated a threat to Europe? In principle yes, only the victorious Macron represents a Europe out of touch with ordinary people, i.e. the very politics which gave such strength to Le Pen, and he is not an outsider but the establishment in its purest form.

Of course Le Pen and Macron are not the same – the difference that separates them is obvious – but nonetheless the choice between the two of them is not a real choice. To see this, it is enough to focus on the background of each candidate: Le Pen is a racist populist, but also addresses popular and workers' dissatisfactions; Macron presents himself as a tolerant, humane pro-European, but the economic policy he stands for is the main cause of popular dissatisfaction with Europe. The sad prospect that awaits us is that of a future in which, every four years, we will be thrown into a panic, scared by some form of 'neo-fascist danger', and in this way be blackmailed into casting our vote for the 'civilized' candidate in meaningless elections lacking any positive vision . . . Meanwhile we'll be able to sleep in the safe embrace of global capitalism with a human face. The obscenity of the situation is breathtaking: global capitalism is now presenting itself as the last protection against fascism; and if you try to point out some of Macron's serious limitations you are accused of – yes, of complicity with fascism, since, as we are told repeatedly by the big (and not so big) media, the extreme Left and extreme Right are now coming together: both are anti-Semitic, nationalist-isolationalist, anti-globalist, etc. This is the point of the whole operation: to make the Left – which means any true alternative – disappear. The woman behind Macron is not his wife but the proverbial TINA – the stance of 'there is no alternative'. Macron doesn't bring hope, he kills hope, the hope that we'll really get rid of the threat of racist populism. And this threat is very real – how it works was made clear on 24 October 2017, when the media reported on the following statements of Viktor

Orbán, the Prime Minister of Hungary, without the letters blushing with embarrassment:

> Orbán called Central and Eastern Europe (CEE) 'zone without migrants'. He claimed this at the celebration of the anniversary of Hungarian Revolution of 1956 on October 23. According to him, the countries of CEE succeed to offer rebuff to the illegal migration and it is the only zone at the European continent that is free from the migrants. 'The mystical financial power stroke Europe with the modern migration, millions of the migrants and invasion of the entrants. This plan was intended to make the mixed continent of Europe but we succeed to offer rebuff. The Polish, Czechs, Slovaks, Romanians and Hungarians should unite in this process,' Orbán claimed. He is sure that every next election in Europe show that the citizens want to decide their way themselves and to take the political situation into their own hands. 'We want safe, fair, Christian and free Europe,' he concluded, and warned: 'We should never underestimate the power of the dark side', referencing *Star Wars* as he referred to the plots of those behind the migrant invasion, adding that they 'have no solid structure but extensive networks'. He also stated: 'The European Union, the European Commission must regain independence from the Soros Empire before the billionaire finishes his program for the destruction of the continent.'[26]

Any association between Orbán's 'zones without migrants' and the old Nazi attempt to create 'zones without Jews' is, of course, purely contiguous. Orbán's reference to the 'dark side' of Europe, the 'mystical financial power' embodied in Soros the Jew, points exactly in this direction, i.e. towards the fascist idea of the plutocrat-Jewish conspiracy. This is how today's extreme populist Right explains the Muslim immigrant 'threat'. In the anti-Semitic imagination, the 'Jew' is the invisible Master who secretly pulls the strings, which is why Muslim immigrants are *not* today's Jews: they are all too visible, not invisible, they are clearly not integrated into our societies, and nobody claims they secretly pull the strings – if one sees in their 'invasion of Europe' a secret plot, then Jews have to be behind it.

The fact that Orbán delivered his speech at the celebration of the anniversary of the Hungarian Revolution of 1956 resonates with unintended ironies. One of the most pathetic moments of the uprising occurred when the Soviet Army was closing in on the rebels, who sent a desperate message to Vienna: 'We are defending the West here.' Now, after the collapse of Communism, the Christian-conservative government portrays as its main enemy Western multicultural consumerist liberal democracy, for which today's Western Europe stands, and calls for a new, more organic communitarian order to replace the 'turbulent' liberal democracy of the last two decades. Orbán has already expressed his approval of 'capitalism with Asian values', so if European pressure on Orbán continues, we can easily imagine him sending the message to the East: 'We are defending Asia here!'[27]

In the US, Trump is defending his own version of 'Asia' (which has nothing to do with the reality of Asia, of course), whose contours are easy to detect. Sometimes, the best way to evaluate a piece of news is to read it alongside another piece of news – such a comparison often enables us to discern the true stakes of a debate. Let's take the reactions to an incisive text: in the summer of 2017, David Wallace-Wells published his essay 'The Uninhabitable Earth',[28] which immediately became famous. It clearly and systematically describes all the threats to our survival, from global warming to the prospect of a billion climate refugees, and the wars and chaos all this will cause. Rather than focusing on the predictable reactions (accusations of scaremongering, etc.), one should read it alongside two facts that are linked to the situation it describes. First, there is, of course, Trump's outright denial of ecological threats; then there is the obscene fact that billionaires who otherwise support Trump are nonetheless getting ready for the apocalypse by investing in luxury underground shelters where they will be able to survive isolated for up to a year, provided with fresh vegetables, fitness centres, and so on.[29]

Another example is a text by Bernie Sanders. In October 2017, Sanders wrote an incisive comment on the Republican budget, the title of which says it all: 'The Republican budget is a gift to billionaires: it's Robin Hood in reverse.'[30] The text is clearly written, full of convincing facts and insights – so why didn't it have more resonance than it did? We should read it alongside media reports about the outrage that exploded when Sanders was announced as an opening-night speaker at the Women's Convention in Detroit. Critics claimed it was bad to let Sanders, a man, speak at a convention devoted to the political advancement of women's rights.[31] No matter that he was to be just one of the two men among sixty speakers, with no transgender representatives (here sexual difference was all of a sudden accepted as unproblematic). Lurking beneath this outrage was, of course, the reaction of the Clinton wing of the Democratic Party to Sanders: its uneasiness with Sanders's Leftist critique of today's global capitalism. When Sanders emphasizes economic problems, he is accused of 'vulgar' class reductionism, while nobody is bothered when leaders of big corporations support LGBT+.

So should we conclude from all this that our task is to depose Trump as soon as possible? When Dan Quayle, not exactly famous for his high IQ, was Vice-President to Bush Senior, a joke was going around that the FBI had a secret order to carry out if Bush died: kill Quayle immediately. Let's hope the FBI has the same order for Pence in the case of Trump's death or impeachment – Pence is, if anything, worse than Trump, a true Christian conservative. What makes the Trump movement minimally interesting are its inconsistencies: recall that Steve Bannon not only opposes Trump's tax plan but openly advocates raising taxes for the rich to 40 per cent, plus he argues that rescuing banks with public money is 'socialism for the rich' – surely not something Pence likes to hear.

Steve Bannon recently declared war, but against whom? Not against Democrats from Wall Street, not against liberal intellectuals or any other usual suspects, but against the Republican Party establishment itself. Having been fired from the White House by Trump, he is fighting for Trump's mission at its purest, even if it is sometimes against Trump himself – let's not forget that Trump is basically destroying the Republican Party. Bannon aims to lead a populist revolt of underprivileged people against the elites – he is taking Trump's message of government by and for the people more literally than Trump himself dares to do. To put it bluntly, Bannon is like the SA with regard to Hitler, the lower-class populist part Trump will have to get rid of (or neutralize at least) in order to be accepted by the establishment and function smoothly as head of state. That's why Bannon is worth his weight in gold: he is a permanent reminder of the antagonism that cuts across the Republican Party. His type of populism is, at least, ready to detect the hypocrisy of the predominnat liberal order. For example, is the basic lesson of the recent public disclosure of the so-called Paradise Papers not the simple fact that the ultra-rich live in their special zones where they are not bound by common laws? Micah White Sums up this lesson in two points:

> First, the people everywhere, regardless of whether they live in Russia or America, are being oppressed by the same minuscule social circle of wealthy elites who unduly control our governments, corporations, universities and culture ... there is a global plutocracy who employ the same handful of companies to hide their money and share more in common with each other than with the citizens of their countries. This sets the stage for a global social movement.
>
> Second, and most importantly, these leaks indicate that our earth has bifurcated into two separate and unequal worlds: one inhabited by 200,000 ultra high-net-worth individuals and the other by the 7 billion left behind.

We don't really learn anything new from the Paradise Papers – we have been vaguely aware of it for a long time. What is new is not that our vague suspicions are now confirmed by precise data, but a change in what, following Hegel, one should call *Sitten* – public customs – which now seem to tolerate much less corruption. One should not idealize this new situation: a fight against corruption can easily be appropriated by conservative anti-liberal forces whose longstanding motto is 'too much democracy brings corruption'. A new space is nonetheless opened up: demanding of the rich and powerful that they obey the laws can be subversive insofar as the system cannot really afford it, i.e. insofar as tax havens and other forms of illegal financial activities are a deeply ingrained part of global capitalism.

The first conclusion we are compelled to draw from this strange predicament is that class struggle is back as the main determining factor of our political life, in the good old Marxist sense of 'determination in the last instance': even if the stakes appear to be totally different, from humanitarian crises to ecological threats, class struggle lurks in the background and casts its ominous shadow.

The second conclusion is that class struggle is less and less directly transposed into the struggle *between* political parties, and increasingly takes place *within* each big political party. In the US, class struggle cuts across the Republican Party (the Party establishment versus Bannon-like populists) and across the Democratic Party (the Clinton wing versus the Sanders movement). We should, of course, never forget that Bannon is the beacon of the alt-right while Clinton supports many progressive causes, such as the fight against racism and sexism. However, at the same time we should never forget that the LGBT+ struggle can also be co-opted by mainstream liberalism against 'class essentialism' of the Left.

The third conclusion thus concerns the Left's strategy in this complex situation. While any pact between Sanders and Bannon is excluded for obvious reasons, a key element of the Left's tactics should be to ruthlessly exploit divisions in the enemy camp and fight for Bannon followers. To cut a long story short, there is no victory of the Left without the broad alliance of all anti-establishment forces. One should never forget that our true enemy is the global capitalist establishment and not the new populist Right, which is merely a reaction to its impasses. If we forget this, then the Left will simply disappear from the map, as is already happening with the moderate Social-Democratic Left in much of Europe (Germany, France . . .), or, as Slawomir Sierakowski put it in his 'In Europe, the Only Choice is Right or Far-Right': 'As left-wing parties have collapsed, the sole option remaining for voters is conservatism or right-wing populism.'[32]

So will Trump get his comeuppance? His impulsive decisions, like his refusal to endorse the G7 declaration agreed upon in Quebec in June 2018, are not just expressions of his personal quirks. They are reactions to the end of an era in the global economic system, reactions that are sustained by a wrong vision of what is going on. However, Trump's wrong vision is nonetheless based on the correct insight that the existing world system no longer works. An economic cycle is coming to an end, a cycle that began in the early 1970s, the time when what Yanis Varoufakis calls the 'Global Minotaur' was born, the monstrous engine that was running the world economy from the early 1980s to 2008. The late 1960s and the early 1970s were not just the times of oil crisis and stagflation; Nixon's decision to abandon the gold standard for the US dollar was the sign of a much more radical shift in the basic functioning of the capitalist system. By the end of the 1960s, the US economy was no longer able to continue the recycling of its surpluses to Europe and Asia: its surpluses had turned into deficits. In 1971, the US government responded to this decline with an audacious strategic move: instead of tackling the nation's burgeoning deficits, it decided to do the opposite, to *boost deficits*. And who would pay for them? The rest of the world! How? By means of a permanent transfer of capital that rushed ceaselessly across the two great oceans to finance America's deficits. These deficits thus started to operate

> like a giant vacuum cleaner, absorbing other people's surplus goods and capital. While that 'arrangement' was the embodiment of the grossest imbalance imaginable at a planetary scale . . . nonetheless, it did give rise to something resembling global balance; an international system of rapidly accelerating asymmetrical financial and trade flows capable of putting on a semblance of stability and steady growth . . . Powered by these deficits, the world's leading surplus economies (e.g. Germany, Japan and, later, China) kept churning out the goods while America absorbed them. Almost 70% of the profits made globally by these countries were then transferred back to the United States, in the form of capital flows to Wall Street. And what did Wall Street do with it? It turned these capital inflows into direct investments, shares, new financial instruments, new and old forms of loans etc.[33]

This growing negative trade balance demonstrates that the US is the non-productive predator: in the last decades, it has had to suck up $1 billion daily from other nations to pay for its consumption and is, as such, the universal Keynesian consumer that keeps the world economy running. (So much for the anti-Keynesian economic ideology that seems to predominate today!) This influx of money, which is effectively like the tithe paid to Rome in Antiquity (or the gifts sacrificed to Minotaur by ancient Greeks), relies on a complex economic mechanism: the US is 'trusted' as the safe and stable centre, so that all others, from the oil-producing Arab countries to Western Europe and Japan, and now even China, invest their surplus profits in the US. Since this trust is primarily ideological and military, not economic, the problem for the US is how to justify its imperial role – it needs a permanent state of war, so it has had to invent the 'war on terror', offering itself as the universal protector of all other 'normal' (not 'rogue') states. The entire globe thus tends to function as a universal Sparta with its three classes, now emerging as the new First, Second, Third world: (1) the US as the military-political-ideological power; (2) Europe and parts of Asia and Latin America as the industrial-manufacturing region (crucial here are Germany and Japan, the world's leading exporters, plus, of course, the rising China); (3) the undeveloped rest, today's helots, those 'left behind'. In other words, global capitalism has brought about a new general trend towards oligarchy, masked as the celebration of the 'diversity of cultures': equality and universalism are more and more disappearing as actual political principles.

From 2008 on, this neo-Spartan world system has been breaking down. In the Obama years, Paul Bernanke, the Chairman of the Federal Reserve, gave another breath of life to this system: exploiting the fact that the US dollar is the global currency, he financed imports by massively printing money. Trump has decided to approach the problem in a different way: ignoring the delicate balance of the global system, he focused on elements that might be presented as 'injustices' for the US: gigantic levels of imports are reducing domestic jobs, and so on. But what he decries as 'injustice' is part of a system that has profited US: the US were effectively 'robbing' the world by importing stuff and paying for it through debts and printing money. And the same game of robbing will obviously go on: Trump did not only lower the taxes for the rich, he also silently endorsed many Democratic demands to alleviate the situation of the poor, which means the deficit will explode ... and when asked about it, Trump will probably repeat the old Reagan answer: 'Our deficit is big enough to take care of itself!'

So it is no surprise that Trump is currently addressing Kim Jong-un in much more friendly terms than his big Western allies: here also, extremes meet, and we should bear in mind that, economically and geopolitically, Europe is Trump's true enemy. With the disintegration of the system that dominated world trade from 1970, the US are more and more becoming the disruptive element of world trade. In contrast to 1945, the world doesn't need the US, it is the US that needs the world. Two outcasts have thus been meeting in Singapore: the excluded outcast (Kim) and the outcast in the very centre of our world. Since Trump has already declared his intention to invite Kim to the White House, I am haunted by a dream – not the noble Martin Luther King one but a much weirder one (much more easy to realize than Luther's dream). Trump has revealed his love for military parades, proposing to organize one in Washington, but the Americans seem not to like the idea. So, what if his new friend Kim can give him a helping hand? What if he returns the invitation and prepares a spectacle for Trump at the big stadium in Pyongyang, with hundreds of thousands of well-trained North Koreans waving colourful flags to form gigantic moving images of Kim and Trump smiling?

In a world where decisions are made in intimate meetings of 'strong leaders', there is no place for Europe as we know it. Obviously, Trump feels most comfortable in the company of authoritarian leaders with whom one can 'make deals' – especially if they act only on behalf of their own state. 'America first' can make a deal with 'China first' or 'Russia first', or the post-Brexit 'UK first', not with a united Europe. Trump's goal is to make trade deals with single partners who can all be blackmailed into submission, so it is of utmost importance that Europe acts as a unified economic and political force. Full of dangers as this new situation is, it opens up a unique chance for Europe: to engage itself in the formation of a new global economic system that will no longer be dominated by the US dollar as the global currency. In the global economy, it's war, so it's time for radical measures. Europe should be aware that there is no return to the status quo perturbed by Trump. A truly new world order is needed for Trump to get his comeuppance. Neither Russia nor China can create this – they are caught in the same game as Trump – they all speak the same language of 'America (Russia, China) first'. Only Europe can do it, but it is here that the reaction of Europe and Canada is insufficient: instead of advocating a new vision, they act as an offended party complaining that the US has broken the established rules. In the last decade or so, the EU has more and more acted like the PLO ex-leader Yasser Arafat, about whom it was said that he never missed an opportunity to miss an opportunity. The immigrant crisis, Catalonia . . . It is probable that Europe will again miss the chance. We Europeans are obviously not strong enough to reject the US order to boycott Iran. (As we know now, immediately after Merkel proudly announced that Germany would act as if the Iran agreement was still on, big German companies silently withdrew from Iran.) At times like this, one is really ashamed to be a European, in spite of all the heroic statements that abound. Here is a proud French reaction to the new situation created by Trump's withdrawal from the Iran nuclear agreement, declaring the European will to assert itself as a sovereign power bloc and to act as if the pact with Iran was still valid:

> Europe is prepared to introduce measures to nullify the effect of Donald Trump imposing sanctions on any non-US firm that continues to do business with Iran, the French government has said. The warning from the French finance minister, Bruno Le Maire, suggests Trump's proposals to corral Europe into joining US foreign policy on Iran may lead to a severe backlash by EU firms and politicians, especially advocates of a stronger independent European foreign policy. 'We have to work among ourselves in Europe to defend our European economic sovereignty,' Le Maire said, adding that Europe could use the same instruments as the US to defend its interests. He added: 'Do we want to be a vassal that obeys and jumps to attention?'[34]

Sounds nice – but does Europe have enough strength and unity to do it? Will the new East European post-Communist 'axis of Evil' (stretching from the Baltic states to Croatia) follow the EU resistance to the US, or will it bow to the US and thus provide yet another proof that the quick expansion of the EU to the East was a mistake? What further complicates things is that Europe is caught in its own populist revolt, triggered by the fact that people trust less and less the Brussels technocracy, experiencing it as a centre of power with no democratic legitimacy. The result of the most recent Italian elections is that, for the first time in a developed Western European country, Euro-sceptic populists have come to power. And remember that the withdrawal from the Iran agreement is just one of three anti-European acts by the US: we also have the moving of the US embassy in Israel from Tel Aviv to Jerusalem, vehemently opposed by the EU, and, of course, the opening shots in a trade war with three of the US's biggest trading partners, levying tariffs on imports of steel and aluminium from the EU, Canada and Mexico.

Although most of us sympathize with the European reaction, we should not forget the (as a rule ignored) background to the US decision. To understand it, let's turn to another topic which may appear to be totally different: the current uproar in the US over the abrupt cancellation of ABC's hit TV show *Roseanne* because of a racist tweet by the show's star, Roseanne Barr. In her column 'With Roseanne Barr gone, will the US working-class be erased from TV?',[35] Joan Williams argues that the Left should finally start to listen to the white working class. She notices how a key fact of this affair passed unnoticed: the cancellation 'deprived American television of one of the only sympathetic depictions of white working-class life in the past half century – in other words, since television began.'[36] Williams unambiguously supports the exclusion of Barr on account of her racist tweets – but she adds: 'All that said, race is not the only social hierarchy. Disrespectful images of the working-class whites are part and parcel of the cultural disrespect that paved the path for a demagogue like Trump.'[37] The sad plight of the working-class whites is the clearest indication of the disappearance of the American dream:

> Virtually all Americans born in the 1940s earned more than their parents; today, *it's less than half*. The rust belt revolt that brought both Brexit and Trump reflects rotting factories, dying towns, and a half century of empty promises. Those left behind are very, very angry; Trump is their middle finger. The more he outrages coastal elites, the more his followers gloat they got our goat. Finally, they are being noticed.[38]

And it is crucial to read Trump's tariff war against the closest allies of the US against this background: in his populist version of class warfare, Trump's goal is (also) to protect the American working class (and are metalworkers not one of the emblematic figures of the traditional working class?) from 'unfair' European competition, thereby saving American jobs. And it is wrong to dismiss Trump as a mere demagogue here:

> Behind the scenes, Trump astonished Nancy Pelosi, the Democrat's leader in the House of Representatives, by approving every single social program that she asked of him . . . Whatever one thinks of this president, he is giving money away not only to the richest, who of course get the most, but also to many poor people.[39]

Again, this is why all the protests of public officials and economists in the EU, Canada and Mexico, as well as the counter-measures proposed by them, miss the target: they follow the WTO logic of free international trade, while only a new Left addressing the concerns of all those left behind can really counter Trump.

At some deep and often obfuscated level, the US neo-cons perceive the European Union as *the* enemy. This perception, kept under control in the public political discourse, explodes in its underground obscene double, the extreme Right Christian fundamentalist political vision, with its obsessive fear of the New World Order (Obama is in secret collusion with the United Nations, international forces will intervene in the US and put all true American patriots in concentration camps – a couple of years ago, there were rumours that Latino American troops were already on the Midwest plains, building concentration camps . . .). One way to resolve this dilemma is the hardline Christian fundamentalist one, articulated in the works of Tim LaHaye *et consortes*: to unambiguously subordinate the second opposition to the first one. The title of one of LaHaye's novels points in this direction: *The Europa Conspiracy*. The true enemies of the US are not Muslim terrorists, they are merely puppets secretly manipulated by the European secularists, the true forces of the anti-Christ who want to weaken the US and establish the New World Order under the domination of the United Nations. In a way, they are right in this perception: Europe is not just another geopolitical power bloc, but a global vision that is ultimately incompatible with nation-states. So what prevents Europe from gathering the strength to strike back?

On 18 May 2017 I had a conversation with Will Self at the Emmanuel Centre in London. Its most memorable moment (for me, at least) occurred when Self – while broadly agreeing with me that, if things continue the way they are now, our society is doomed, and an unthinkable catastrophe lies ahead – reproached me for still counting on some big 'revolutionary' act that will turn the global tide and prevent our slide towards catastrophe. His main reason was that, with our way of life, we are so deeply immersed in the process of (not only ecological) self-destruction that no awareness of what we are doing can stop us doing it. Self then asked the public how many of them had smartphones, reminding them that each phone needs coltan, a precious metal from the Congo, where it is mined by *de facto* slave labour in a way detrimental to the environment. So what can we do, having admitted that we are all co-responsible and unable to actively intervene? Self's answer: nothing big, just pay our taxes (to enable the state to maintain minimal order of law and welfare) and enjoy our isolated life, jerking off . . . My reply (which I failed to articulate properly there) is that such a cynical-hedonist stance perfectly suits those in power, that it is ideology at its purest: any collective *counter*-act is in advance devalued ('Who are you to protest? Are you not also using coltan? So what right do you have to put the blame on big corporations?'), so all we can do is remain private citizens who masochistically enjoy our guilt and withdraw into private pleasures. This state of things does not mean that we are lost; it rather points in the direction of Joel 3:14 – 'Multitudes, multitudes in the valley of decision: for the day of the Lord is near in the valley of decision.' These lines provide the first accurate description of the moment when a society is at a crossroads, confronted with a choice that may decide its fate. This is the situation of Europe today.

Every anti-immigrant populist would fully agree with this claim: yes, Europe's very identity is threatened by the invasion of Muslim and other refugee multitudes. But the actual situation is exactly the opposite: it is today's anti-immigrant populists who are the true threat to the emancipatory core of the European Enlightenment. A Europe where Marine Le Pen or Geert Wilders are in power is no longer Europe. So what is this Europe worth fighting for?

The true novelty of the French Revolution resides in the distinction between citizens' rights and human rights. One should reject here the classic Marxist notion of human rights as the rights of the members of bourgeois civil society. While citizens are defined by the political order of a sovereign state, 'human' is what remains of a citizen when he or she is deprived of citizenship, finding him- or herself in what in artillery one calls the open space, reduced to an abstract talking body. It is in this sense that universal human rights should remain our point of reference when we negotiate the difficult relationship between the constraints of citizenship and particular ways of life. Without this compass we inevitably regress to barbarism.

In his reading of the (in)famous difference between human rights and citizens' rights, Milner rejects the Marxist critical notion of human rights as those of members of bourgeois civil society: for Milner, a citizen is a member of a community, sharing its specific culture, while a human being is what remains of a citizen when he or she is deprived of his or her citizenship. Human rights are 'natural' rights only in this sense of externality to a particular culture; they have nothing to do with eternal nature, since they apply to what remains of a citizen after she or he is subtracted from a specific polis. In this sense, their 'nature' is a retroactive effect of culture; it applies to a human being reduced to the zero level of a speaking body:

> one gains a glimpse into the real of the rights of the body in examining what goes on when they are denied to individuals. Every day brings us a new example. I do not have to think about bombs and poisonous gasses, I think about Calais: those who are assembled there from 2000 are not guilty of anything, they are not accused of anything, they do not infringe upon any part of the law; they are simply there and they live; the proof that they live is that sometimes they die. Nobody knows what languages they are speaking and anyway one doesn't listen to them. One only knows that they speak. They are therefore reduced to the status of speaking bodies; by the settlement to which they are submitted they literally render visible in a negative way the real of the rights of man/woman . . . These rights are openly distinguished from the rights of a citizen since refugees are precisely not the citizens of Calais and mostly do not want to become that.[40]

Milner insists on the 'vulgar' materiality of these rights:[41] they are more basic than the right to organize public meetings, to practise free speech, to freely express opinions, and so on. Before that comes the material needs of a body: water, food, hygiene, minimal space for privacy. If individuals are deprived of this, their 'higher' human rights disappear. Human rights are, first, basic material rights: toilets, kitchens, healthcare. Rights begin with the space for secretion – this is the sad basis of my story about the different shapes of European toilets. Insofar as human rights (as distinct from citizens' rights) were first proclaimed during the French Revolution, one should note the irony of the fact that Calais is a French city. Here, of course, we enter a double game: Marxists emphasize 'material' rights against freedom of opinion, freedom of the press, etc. (but fail to deliver them when in power), while 'bourgeois democracies' emphasize other freedoms.

The lesson here is that universal human rights are – in their very universality – historically produced and specified; their exact extent and content is the result of socio-political struggles. Is Milner's renaming of them as 'the rights of man/woman' not itself the effect of contemporary feminist struggles? Plus we should bear in mind that, although humans who are covered by these rights are 'proletarian' in the sense of being deprived of citizenship, they are nonetheless not abstract Cartesian cogitos – they come as individuals embedded in a specific way of life often in conflict with that of the country in which they dwell as refugees. So we have to take into account three levels here: the abstract universality of a human being *qua* bearer of human rights, the particularity of a specific way of life to which an individual belongs, and the singularity of citizenship as the mediating moment between the two extremes (as a citizen, I am universal, but universal as belonging to the singularity of a state). The interaction of these three levels cannot but engender multiple difficulties – suffice it to recall the vagaries of power that plague contemporary attempts to enact radical emancipation.

In order to deal with these difficulties, another triad of the universal, the particular, and the singular should be brought into play: the triad of (universal) mass uprising, the (particular) political organization, and . . . what should stand for the singular? This third element brings us back to Lenin, with whom we began the chapter. In the overflow of celebratory reactions to the centenary of the October Revolution in 2017, its central lesson for today passed unnoticed (or was mentioned as a proof that the October Revolution was a coup performed by a secretive group and not a true popular uprising at all). This lesson concerns the unique collaboration between Lenin and Trotsky.

The kernel of Lenin's 'utopia' arises out of the ashes of the catastrophe of 1914, in his settling of accounts with the Second International orthodoxy: the radical imperative to smash the bourgeois state, which means the state *as such*, and to invent a new communal social form without a standing army, police or bureaucracy, in which all could take part in the administration of social matters. This was for Lenin no theoretical project for some distant future – in October 1917, Lenin claimed that 'we can at once set in motion a state apparatus constituting of ten if not twenty million people.'[42] *This urge of the moment is the true utopia.* What one should stick to is the *madness* (in the strict Kierkegaardian sense) of this Leninist utopia – if anything, Stalinism stands for a return to the realistic 'common sense'.

One cannot overestimate the explosive potential of *The State and Revolution* – in this book, 'the vocabulary and grammar of the Western tradition of politics was abruptly dispensed with.'[43] What then followed can be called, borrowing the title of Althusser's text on Machiavelli, *La solitude de Lénine*: the time when he basically stood alone, struggling against the current in his own party. When, in his 'April Theses' of 1917, Lenin discerned the *Augenblick*, the unique chance for a revolution, his proposals were first met with stupor or contempt by a large majority of his party colleagues. Within the Bolshevik party, no prominent leader supported his call to revolution, and *Pravda* took the extraordinary step of dissociating the party, and the editorial board as a whole, from Lenin's 'April Theses' – far from being an opportunist flattering and exploiting of the prevailing mood of the populace, Lenin's views were highly idiosyncratic. Bogdanov characterized the 'April Theses' as 'the delirium of a madman',[44] and even Nadezhda Krupskaya, Lenin's wife, concluded that 'I am afraid it looks as if Lenin has gone crazy.'[45]

In February 1917 Lenin was stranded in Zurich, with no reliable contacts in Russia, learning about events there mostly from the Swiss press; in October, he led the first successful socialist revolution – so what happened in between? In February, Lenin immediately perceived the revolutionary chance, the result of unique contingent circumstances – if the moment wasn't seized, the chance for revolution would be forfeited, perhaps for decades. Even a couple of days before the October Revolution, Lenin wrote: 'The triumph of both the Russian and the world-revolution depends on a two or three days' struggle.' In his stubborn insistence that one should take the risk and pass to the act, Lenin was alone, ridiculed by the majority of the Central Committee members of his own party. However, indispensable as Lenin's personal intervention was, one should not modify the story of the October Revolution into that of the lone genius confronted with the disoriented masses and gradually imposing his vision. Lenin succeeded because his appeal, while bypassing the party *nomenklatura*, found an echo in what one is tempted to call revolutionary micropolitics: the incredible explosion of grassroots democracy, of local committees popping up all around Russia's big cities and, while ignoring the authority of the 'legitimate' government, taking things into their hands. This is the untold story of the October Revolution, the obverse of the myth of the tiny group of ruthless, dedicated revolutionaries accomplishing a *coup d'état* . . .

On the other hand, the notion that a tiny group of ruthless, dedicated revolutionaries accomplished a *coup d'état* is not *just* a myth – there is a crucial grain of truth in it. When popular dissatisfaction grew and Lenin's idea that a revolutionary situation was emerging began to be accepted, the majority of the Bolshevik party leaders wanted to organize a mass popular uprising. Trotsky, however, advocated a view that, to traditional Marxists, couldn't but appear as 'Blanquist': a narrow, well-trained elite should take power. After a short oscillation, Lenin defended Trotsky, specifying why Trotsky is not advocating Blanquism:

> In [a] letter of October 17, Lenin defended Trotsky's tactics: 'Trotsky is not playing with the ideas of Blanqui,' he said. 'A military conspiracy is a game of that sort only if it is not organized by the political party of a definite class of people and if the organizers disregard the general political situation and the international situation in particular. There is a great difference between a military conspiracy,

which is deplorable from every point of view, and the art of armed insurrection.'[46]

In this precise sense, 'Lenin was the "strategus", idealist, inspirer, the *deus ex machina* of the revolution, but the man who invented the technique of the Bolshevik *coup d'état* was Trotsky.'[47] Against later 'Trotskyite' defenders of an (almost) 'democratic' Trotsky who advocates authentic mass mobilization and grass-roots democracy, one should emphasize that Trotsky was all too aware of the inertia of the masses – the most one can expect of the 'masses' is chaotic dissatisfaction. A tightly defined, well-trained revolutionary force should use this chaos to strike at power and thereby open up the space in which the masses can really organize themselves . . . Here, however, the crucial question arises: what does this narrow elite do? In what sense does it 'take power'? The true novelty of Trotsky becomes visible here: the striking force does not 'take power' in the traditional sense of a palace *coup d'état*, occupying government offices and army headquarters, it does not focus on confronting the police or the army on the barricades. Let us quote some passages from Curzio Malaparte's unique *The Technique of Coup d'État* (1931) to get the taste of it:

> Kerenski's police and the military authorities were especially concerned with the defence of the State's official and political organizations: the Government offices, the Maria Palace where the Republican council sat, the Tauride Palace, seat of the Duma, the Winter Palace, and General Headquarters. When Trotsky discovered this mistake he decided to attack only the technical branches of the national and municipal Government. Insurrection for him was only a question of technique. 'In order to overthrow the modern State,' he said, 'you need a storming party, technical experts and gangs of armed men led by engineers.'
>
> While Trotsky was organizing the *coup d'état* on a rational basis, the Central Committee of the Bolshevik Party was busy organizing the proletarian revolution. Stalin, Sverdlov, Boubrov, Ouritzki, and Dzerjinski, the members of this committee who were developing the plan of the general revolt, were nearly all openly hostile to Trotsky. These men felt no confidence in the insurrection as Trotsky planned

> it, and ten years later Stalin gave them all the credit for the October *coup d'état*.
>
> On the eve of the *coup d'état*, Trotsky told Dzerjinski that Kerenski's government must be completely ignored by the Red Guards; that the chief thing was to capture the State and not to fight the Government with machine-guns; that the Republican Council, the Ministries and the Duma played an unimportant part in the tactics of insurrection and should not be the objectives of an armed rebellion; that the key to the State lay, not in its political and secretarial organizations nor yet in the Tauride, Maria or Winter Palaces, but in its technical services, such as the electric stations, the telephone and telegraph offices, the port, gasworks and water mains.[7]

Trotsky thus targeted the material (technical) grid of power (railways, electricity, water supply, post, etc.), the grid without which state power hangs in the void and becomes inoperative. Let the mobilized masses fight the police and storm the Winter Palace (an act without any real relevance): the essential move is accomplished by a tiny, dedicated minority . . .

Instead of indulging in a miserable moralist-democratic rejection of such a procedure, one should rather analyse it coldly and think about how to apply it today, since this insight of Trotsky has gained new actuality with the progressive digitalization of our lives in what could be characterized as the new era of posthuman power. Most of our activities (and passivities) are now registered in some digital cloud that also permanently evaluates us, tracing not only our acts but also our emotional states; when we experience ourselves as free to the utmost (surfing the web, where everything is available), we are totally 'externalized' and subtly manipulated. The digital network gives new meaning to the old slogan 'the personal is political'. And it's not only the control of our intimate lives that is at stake: everything is today regulated by some digital network, from transport to health, from electricity to water. That's why the web is now our most important commons, and the struggle for its control is *the* struggle today. The enemy is the combination of privatized and state-controlled commons, corporations (Google, Facebook) and state security agencies (NSA). But we know all this, so where does Trotsky enter?

The digital network that sustains the functioning of our societies as well as their control mechanisms is the ultimate figure of the technical grid that sustains power – and does this not confer a new lease of life on Trotsky's idea that the key to the State lies not in its political and secretarial organizations but in its technical services? Consequently, in the same way that, for Trotsky, taking control of the post, electricity, railways and so on was the key moment of the revolutionary seizure of power, is it not the case that today the 'occupation' of the digital grid is absolutely crucial if we are to break the power of the state and capital? And, in the same way that Trotsky required the mobilization of a tight, disciplined 'storming party, technical experts and gangs of armed men led by engineers' to resolve this 'question of technique', the lesson of the last decades is that neither massive grass-roots protests (as we have seen in Spain and Greece) nor well-organized political movements (parties with elaborated political visions) are enough – we also need a narrow, striking force of dedicated 'engineers' (hackers, whistle-blowers . . .) organized as a disciplined conspiratorial group. Its task will be to 'take over' the digital grid, to rip it out of the hands of corporations and state agencies that now *de facto* control it.

WikiLeaks was here just the beginning, and our motto should be here a Maoist one: Let a hundred WikiLeaks blossom! The panic and fury with which those in power, those who control our digital commons, reacted to Assange is a proof that such an activity hits the nerve. There will be many blows below the belt in this fight – our side will be accused of playing into the enemy's hands (like the campaign against Assange for being in the service of Putin), but we should get used to it and learn to strike back with interest, ruthlessly playing one side against another in order to bring them all down. Were Lenin and Trotsky also not accused of being paid by the Germans and/or by Jewish bankers? As for the scare that such an activity will disturb the functioning of our societies and thus threaten millions of lives: we should bear in mind that it is those in power who are ready to selectively shut down the digital grid to isolate and contain protests – when massive public dissatisfactions explode, the first move is always to disconnect the internet and mobile phones.

We need thus the political equivalent of the Hegelian triad of the universal, the particular, and the singular. Universal: a mass upheaval, in the Podemos style. Particular: a political organization that can translate the dissatisfaction into an operative political programme. Singular: 'elitist' specialized groups which, acting in a purely 'technical' way, undermine the functioning of state control and regulation. Without this third element, the first two remain impotent.

3
From Identity to Universality

What Agatha Knew

Agatha Christie's eightieth book, *Passenger to Frankfurt* – published in 1970 with the subtitle 'an extravaganza', one of the few of her works with no movie or TV adaptation – is a novel that 'slides from the unlikely to the inconceivable and finally lands up in incomprehensible muddle. Prizes should be offered to readers who can explain the ending. Concerns the youth uproar of the 'sixties, drugs, a new Aryan superman and so on, subjects of which Christie's grasp was, to say the least, uncertain.'[1] However, this 'incomprehensible muddle' is not due to Christie's senility: its causes are clearly political. *Passenger to Frankfurt* is Christie's most personal, intimately felt and at the same time most political novel; it expresses her personal confusion, her feeling of being totally at a loss with what was going on in the world in the late 1960s – drugs, sexual revolution, student protests, murders, etc. It is crucial to note that this overwhelming feeling of confusion is formulated by an author whose speciality was detective novels, stories about crime, stories about the darkest side of human nature. The deeper reason for her despair is the feeling that, in the chaotic world of 1970, it was no longer possible to write detective novels that still presupposed a stable society based on law and order, momentarily disturbed by crime but restored to order by the detective. In the society of 1970, chaos and crime were rife, so no wonder *Passenger to Frankfurt* is not a detective novel: there is no murder, no logic, no deduction. Christie's sense of the collapse of the elementary cognitive mapping, her overwhelming fear of chaos, is rendered clearly in her Introduction to the novel:

> It is what the Press brings to you every day, served up in your morning paper under the general heading of News. Collect it from the front page. What is going on in the world today? What is everyone saying, thinking, doing? Hold up a mirror to 1970 in England.
>
> Look at that front page every day for a month, make notes, consider and classify.
>
> Every day there is a killing.
>
> A girl strangled.
>
> Elderly woman attacked and robbed of her meagre savings.
>
> Young men or boys – attacking or attacked.
>
> Buildings and telephone kiosks smashed and gutted.
>
> Drug smuggling.
>
> Robbery and assault.
>
> Children missing and children's murdered bodies found not far from their homes.
>
> Can this be England? Is England *really* like this? One feels – no – not yet, *but it could be.*
>
> Fear is awakening – fear of what may be. Not so much because of actual happenings but because of the possible causes behind them. Some known, some unknown, but *felt*. And not only in our own country. There are smaller paragraphs on other pages – giving news from Europe – from Asia – from the Americas – Worldwide News.
>
> Hi-jacking of planes.
>
> Kidnapping.
>
> Violence.
>
> Riots.
>
> Hate.
>
> Anarchy – all growing stronger.
>
> All seeming to lead to worship of destruction, pleasure in cruelty.
>
> What does it all mean?

So what *does* all this mean? In the novel, Christie provides an answer; here is the storyline. On a flight home from Malaya, Sir Stafford Nye, a bored diplomat, is approached in the passenger lounge at Frankfurt airport by a woman whose life is in danger; to help her, he agrees to lend her his passport and boarding ticket. In this way, he unwittingly gets caught up in an international intrigue from which the only escape is to outwit the power-crazed Countess von Waldsausen, who wants to achieve world domination by manipulating and arming the planet's youth. This terrible worldwide conspiracy has something to do with Richard Wagner and 'The Young Siegfried'. We learn that, towards the end of the Second World War, Hitler went to a mental institution, met with a group of people who thought they were Hitler, and exchanged places with one of them, thus surviving the war. He then escaped to Argentina, where he married and had a son who was branded with a swastika on his heel: 'The Young Siegfried'. Meanwhile, in the present, drugs, promiscuity and student protests are all secretly caused by Nazi agitators who want to bring about anarchy so that they can restore Nazi domination on a global scale.

This 'terrible worldwide conspiracy' is, of course, ideological fantasy at its purest: a weird condensation of the fear of extreme Right and extreme Left. The least we can say in Christie's favour is that she locates the heart of the conspiracy in the extreme Right (neo-Nazis), not any of the other usual suspects (Communists, Jews, Muslims). The idea that neo-Nazis were behind the '68 student protests and the struggle for sexual liberation, with its obvious madness, nonetheless bears witness to the disintegration of a consistent cognitive mapping of our predicament; the fact that Christie is compelled to take refuge in such a crazy paranoiac construct indicates the utter confusion and panic in which she found herself. The picture of our society she paints is simply confused, out of touch with reality (incidentally, although to a much lesser extent, the same goes for the strangest of John le Carré's novels, *A Small Town in Germany*, which is set in a similar situation). But is her vision really too crazy to be taken seriously? Is our era, with 'leaders' like Donald Trump and Kim Jong-un, not as crazy as her vision? Are we today not all like a bunch of passengers to Frankfurt? Our situation is messy in a way that is very similar to the one described by Christie: we have a Rightist government enforcing workers' rights (in Poland), a Leftist government pursuing the strictest austerity politics (in Greece). No wonder that, in order to regain a minimal cognitive mapping, Christie resorts to the Second World War, 'the last good war', retranslating our mess into its coordinates.

One should nonetheless note how the very form of Christie's denouement (one big Nazi plot behind it all) strangely mirrors the fascist idea of the Jewish conspiracy. Today, the extreme populist Right proposes a similar explanation of the Muslim immigrant 'threat'. In the anti-Semitic imagination, the 'Jew' is the invisible Master who secretly pulls the strings – which is precisely why Muslim immigrants are *not* today's Jews: they are all too visible, not invisible, they are clearly not integrated into our society, and nobody claims they secretly pull any strings. If one sees in their 'invasion of Europe' a secret plot, then Jews have to be behind it – as was suggested in an article that recently appeared in one of the main Slovene Rightist weekly journals, which read: 'George Soros is one of the most depraved and dangerous people of our time,' responsible for 'the invasion of the negroid and Semitic hordes and thereby for the twilight of the EU . . . as a typical talmudo-Zionist, he is a deadly enemy of Western civilization, nation state and white, European man.' His goal is to build a 'rainbow coalition composed of social marginals like faggots, feminists, Muslims and work-hating cultural Marxists', which would then perform 'a deconstruction of the nation state, and transform the EU into a multicultural dystopia of the United States of Europe'. Furthermore, Soros is inconsistent in his promotion of multiculturalism:

> He promotes it exclusively in Europe and the USA, while in the case of Israel, he, in a way which is for me totally justified, agrees with its monoculturalism, latent racism and building of a wall. In contrast to EU and USA, he also does not demand that Israel open its borders and accept 'refugees'. A hypocrisy proper to Talmudo-Zionism.[2]

Is this disgusting fantasy, which brings together anti-Semitism and Islamophobia, so different from that contrived by Christie? Are they both not a desperate attempt to orient oneself in confused times? The extreme oscillations in public perception of the Korean crisis are significant as such. One week we are told that we are on the brink of nuclear war, then there is seven days' respite, then the war threat explodes again. When I visited Seoul in August 2017, my friends there told me there was no serious threat of a war since the North Korean regime knows it could not survive it, now that the South Korean authorities are preparing the population for a nuclear war. Not long ago our media was reporting on the increasingly ridiculous exchange of insults between Kim Jong-un and Donald Trump. The irony was that, in a situation where two apparently immature men are getting angry and hurling insults at each other, our only hope is for some anonymous and invisible institutional constraint to prevent their rage from exploding into full war. Usually we tend to complain that in today's alienated and bureaucratized politics, institutional pressures and constraints prevent politicians from expressing their real point of view; now we hope that such constraints will prevent the expression of all-too-crazy personal visions. How did we reach this point?

Alain Badiou recently warned about the dangers of the growing post-patriarchal nihilist order that presents itself as the domain of new freedoms. The disintegration of the shared ethical basis of our lives is clearly signalled by the abolition of universal military conscription in many developed countries: the very notion of being ready to risk one's life for a common-cause army appears more and more pointless, if not directly ridiculous, so that the armed forces, as the body in which all citizens equally participate, are gradually becoming a mercenary force. This disintegration affects the two sexes differently: men are slowly turning into perpetual adolescents, with no clear passage of initiation into maturity (military service, acquiring a profession, even education no longer play this role). No wonder, then, that, in order to supplant this lack, post-patriarchal youth gangs proliferate, providing ersatz-initiation and social identity. In contrast to men, women today are more and more precociously mature, expected to control their lives, to plan their careers; in this new version of sexual difference, men are ludic adolescents, outlaws, while women appear hard, mature, serious, legal and punitive. A new feminine figure is thus emerging: a cold, competitive agent of power, seductive and manipulative, attesting to the paradox that 'in the conditions of capitalism women can do better than men' (Badiou): contemporary capitalism has invented its own ideal image of woman.

This brings us back to Trump and Kim, these two eternal adolescents, both prone to irrational, brutal outbursts that hurt their own chances. Although the differences between North Korea and the US are obvious, one should nonetheless insist that they both cling to the extreme version of state sovereignty ('Korea first'; 'America first'), plus that the obvious madness of North Korea (a small country ready to risk it all and bomb the US) has its counterpart in the US, which is still pretending to play the role of the global policeman, a single state assuming the right to decide which other states should be allowed to possess nuclear weapons. The solution is thus not to crush North Korea but to find a genuine way of 'internationalizing' nuclear weapons, to make the situation in which a single sovereign state is allowed to possess them (and threaten others with them) unacceptable. The moment we focus just on the 'madness' of North Korea, we already endorse the premise that it should not be allowed to do what only the selected 'superpowers' can do – we should strive to change the entire situation.

This urge to change the entire order emerges precisely when we are facing a threat of total destruction (by nuclear war or ecological catastrophe). In such a situation, our first reaction is a defensive effort to guarantee our survival: let's forget about big emancipatory projects of radical change, our task is now to fight for the survival of what we have, with all the compromises and moderation such an undertaking involves . . . but what is it that we have? The threat of the total destruction of humanity makes us aware of the totality of humanity – humanity appears as one entity only against the background of its (self-)destruction, it was not visible as such before. So the true choice is not between keeping what we have and losing it all (or, in Cold War terms, between developing nuclear arms to protect our liberal freedoms and abandoning nuclear arms but exposing ourselves to the risk of losing our freedoms); as Alenka Zupančič put it:

> the true choice is between losing it all and *creating what we are about to lose*: only this could eventually save us, in a profound sense . . . When caught in the threat and fear of 'losing it all' we are in fact held hostages of something that does not exist (yet). And is this kind of blackmail not actually the very means of making sure that it *never will exist*? It makes us focus on preserving what is there, and what we have, but excludes any real alternative, and means of really thinking differently . . . The possible awakening call of the bomb is not simply 'let's do all in our power to prevent it before it is too late,' but rather 'let's first build this totality (unity, community, freedom) that we are about to lose through the bomb'.[3]

Therein resides the unique chance opened up by the very real threat of nuclear (or ecological, for that matter) destruction: when we become aware of the danger that we might lose it all, we automatically get caught up in a retroactive illusion, a short circuit between reality and its hidden potentials. What we want to save is not the reality of our world, but the reality *as it might have been if it were not hindered by antagonisms that gave birth to the nuclear threat.* This is our true choice when we are confronting total destruction: that between a panicky retreat into self-preservation and active engagement for a change which aims at much more. If we muster the strength for the second choice, then – in Hegelian terms – we pass from 'abstract universality' (of the unleashed negativity that can only achieve global nuclear destruction) to 'concrete universality' (of an alternative new order in which such catastrophes will no longer be possible).

What is needed is no less than a new and global anti-nuclear movement, a global mobilization that would exert pressure on nuclear powers and act aggressively, organizing mass protests, boycotts, etc. It should focus not only on North Korea but also on those superpowers who assume the right to monopolize nuclear weapons. Public mention of the use of nuclear weapons should be treated as a criminal offence, and leaders who openly display their readiness to imperil millions of innocent lives in order to protect their own power should be treated as the worst criminals.

How to Fight Huntington's Disease

It is significant that Trump's first foreign trip was to Saudi Arabia and Israel. If we combine this with his triumphant reception of President el-Sisi of Egypt in the White House, we can see how a new Middle East axis of evil is taking form with full US support: Turkey, Saudi Arabia, Israel, Egypt. The latest brutal exclusion of Qatar is the first big act of this axis, probably a punishment for the positive role of Al Jazeera in the Arab Spring. The breathtaking irony is that it is done in the name of the fight against terrorism, while Saudi Arabia is engaged in the most repressive state terror in Yemen, bombing and displacing millions. The fact that this state terror is more or less ignored by our big media says it all – deplorable as the latest terrorist attacks in London are, we should point out that those who don't want to talk about Yemen should also keep silent about the attacks in London and Paris.

It is the geopolitical background to these tectonic shifts that should worry us. A cartoon published back in July 2008 in the Viennese daily *Die Presse* depicted two stocky Nazi-looking Austrians sitting at a table; one of them holds in his hands a newspaper and comments to his friend: 'Here you can see again how a totally justified anti-Semitism is being misused as a cheap critique of Israel!' This caricature inverts the standard argument against the critics of the policies of the State of Israel, and when today's Christian fundamentalist supporters of Israeli politics reject Leftist critiques of Israeli policies, is their implicit line of argumentation not uncannily close to its reasoning? Remember Anders Breivik, the Norwegian anti-immigrant mass murderer: he was anti-Semitic but pro-Israel, since he saw in the State of Israel the first line of defence against Muslim expansion. He even wants to see the Jerusalem Temple rebuilt, but he wrote in his 'Manifesto': 'There is no Jewish problem in Western Europe (with the exception of the UK and France) as we only have 1 million in Western Europe, whereas 800,000 out of these 1 million live in France and the UK. The US on the other hand, with more than 6 million Jews (600% more than Europe) actually has a considerable Jewish problem.' Breivik thus realizes the ultimate paradox of the Zionist anti-Semite.[4]

The pseudo-Leftist counterpart to this paradox is best exemplified by a graffiti on a wall in Ljubljana, my home city: 'If I were a Palestinian from the West Bank, I would also be a Holocaust denier' – this is exactly the logic one should avoid at any cost, if for no other reason than it reproduces the Zionist argument: 'A Holocaust survivor has the right to ignore minor injustices the State of Israel is committing against Palestinians.' In both cases, one's victimhood is used as the justification for the racist treatment of one's opponents, in line with the reasoning, 'One should understand occasional anti-Semitic outbursts among Arabs in view of Palestinian suffering, and one should understand Israeli politics in the West Bank in view of the horrible past of anti-Semitic pogroms.' Such reasoning is nothing short of blasphemy: on the Zionist side, it reduces the unimaginable horror of the Holocaust to an instrument of local politics and is thereby an affront to the millions of its victims. The only proper ethical stance here is one of universal solidarity: we should support the Palestinian fight for autonomy, not in spite of occasional Arab anti-Semitism but for the same reason we should remember the Holocaust. The pseudo-anti-imperialist formula of Zionism as the exemplary form of today's racism is wrong in exactly the same way as the Zionist formula of 'anti-Semitism and other forms of racism', which tends to disqualify every critique of the State of Israel as anti-Semitic. Every hypocrisy should be rejected here – there should be no 'understanding' for the aggressive way in which local Jews are often mistreated by the Muslim population in Norway or Sweden, there should be no 'understanding' for the way women and gays are treated among many Muslim groups and in many Muslim states. The alliance between the Western radical Left and 'anti-imperialist' fundamentalist Muslims, which makes politically correct Western radicals and Muslim fundamentalists strange bedfellows, is to be rejected as an ideological abomination. When the fight against anti-Semitism and the struggle for Palestinian rights are not conceived as parts of the same endeavour, we find ourselves in a new state of barbarism.

Today we are seeing a novel version of this Zionist anti-Semitism: *Islamophobic respect for Islam*. The same politicians who warn of the danger of the Islamization of the Christian West, from Trump to Putin, respectfully congratulated Erdoğan for his victory in the Turkish referendum (which granted him sweeping new powers) – the authoritarian reign of Islam is OK for Turkey but not for us.[5] So we can well imagine a new version of the caricature from *Die Presse* with two stocky Nazi-looking Austrians sitting at a table, one of them holding in his hands a newspaper and commenting to his friend: 'Here you can see again how a totally justified Islamophobia is being misused for a cheap critique of Turkey!' How are we to understand this weird logic? It is a reaction, a false cure, to the big social disease of our time: Huntington's.

The typical first symptoms of Huntington's disease are jerky, random and uncontrollable movements called chorea, which may initially present as general restlessness, small, unintentional or uncompleted motions, lack of coordination . . . Does an explosion of brutal populism not look quite similar? It begins with what appears to be random, violent excesses against immigrants, outbursts which lack organization and just express a general unease and restlessness apropos 'foreign intruders', but then it gradually grows into a well-coordinated and ideologically grounded movement – what the other Huntington (Samuel) called 'the clash of civilizations'. This coincidence is telltale: what is usually referred to under this term is effectively the Huntington's disease of today's global capitalism.

According to Samuel Huntington, since the end of the Cold War the 'iron curtain of ideology' has been replaced by the 'velvet curtain of culture'. Huntington's dark vision of the 'clash of civilizations' may appear to be the very opposite of Francis Fukuyama's bright prospect of the End of History in the guise of a worldwide liberal democracy. What can be more different from Fukuyama's pseudo-Hegelian idea that the final formula of the best possible social order was found in capitalist liberal democracy than a 'clash of civilizations' as the main political struggle in the twenty-first century? How, then, do the two fit together?

From today's experience, the answer is clear: the 'clash of civilizations' *is* politics at 'the end of history'. The ethnic-religious conflicts are the form of struggle which fits global capitalism: in our age of 'post-politics', when politics proper is progressively replaced by expert social administration, the only remaining legitimate source of conflict is cultural (ethnic, religious) tensions. Today's rise of 'irrational' violence is thus to be conceived as strictly correlative to the depoliticization of our society, i.e. to the disappearance of the proper political dimension and its translation into different levels of 'administration' of social affairs. If we accept this thesis, the only alternative to the 'clash of civilizations' remains the peaceful coexistence of civilizations (or of 'ways of life', a more popular term today): forced marriages and homophobia (or the idea that a woman going alone to a public place is asking to be raped) are OK, it's just that they are limited to another country which is otherwise economically fully included in the world market.

The New World Order that is emerging is thus no longer the Fukuyamaist one of global liberal democracy, but that of the peaceful coexistence of different politico-theological ways of life – coexistence, of course, in the context of the smooth functioning of global capitalism. The obscenity of this process is that it can present itself as progress in anti-colonial struggle: the liberal West will no longer be allowed to impose standards on others, all ways of life will be treated as equal . . . no wonder Robert Mugabe displayed sympathy for Trump's slogan 'America first!': 'America first!' for you, 'Zimbabwe first!' for me, 'India first!' or 'North Korea first!' for them. This is how the British Empire, the first global capitalist empire, functioned: each ethnic-religious community was allowed to pursue its own way of life – Hindus in India were safely burning widows, and so on – and these local 'customs' were either criticized as barbaric or praised for their premodern wisdom, but tolerated since what mattered was that economically they were part of the Empire.

Trump's 'America first' politics opened up space for China to present itself as the agent of a new globalization: in the autumn of 2017 the Chinese President, Xi Jinping, urged world leaders to reject protectionism, embrace globalization and pull together 'like an airborne skein of long-necked geese'. Xi celebrated the new Chinese multi-billion-dollar infrastructure projects as a means of building a modern-day version of the ancient Silk Road, announcing a new golden age of globalization. It is crucial to see that there is no contradiction between market globalization and the accent on one's own particular 'way of life' in the cultural sphere.

This is why the recent display of new German–French unity, as well as Angela Merkel's statement that Europe will have to learn to stand on its own and stop relying on the US for protection, is a more than welcome sign of growing European self-awareness: there is no place for what Europe stands for in the New World Order of Trump, Putin, Modi, Mugabe and Erdoğan. Europe will have to assert its emancipatory legacy by fighting not these foreign threats, but its own version of them – the threat of nationalist populism. And this is why the idea of European union is worth fighting for, in spite of the misery of its actual existence: in today's global-capitalist world, it offers the only model of transnational organization with the authority to limit national sovereignty and the task of guaranteeing a minimum of ecological and social-welfare standards. Something that descends directly from the best traditions of the European Enlightenment survives in it. Our duty is not to humiliate ourselves as the ultimate culprits of colonialist exploitation, but to fight for this part of our legacy, which is important for the survival of humanity. Europe is more and more alone in the new global world, dismissed as an old, exhausted, irrelevant continent playing a secondary role in today's geopolitical conflicts. As Bruno Latour recently put it: 'L'Europe est seule, oui, mais seule l'Europe peut nous sauver' (Europe is alone, yes, but Europe alone can save us).

One of the reliable signs of political opportunism is what, in parallel with particle physics, one may call political correlationism. Let's say I and my enemy both hold in our hand a ball which can be either white or black, and neither of us knows its colour. I am not allowed to look into my own folded palm, so we have here four possibilities: white-white, black-black, black-white and white-black. Now let's suppose that, for some reason, we both know that the two balls (the one in my hand and the one in my enemy's hand) are different in colour; in this case there are only two possibilities, black-white and white-black; and if by chance I get to know the colour of the ball in my enemy's hand, I automatically know the colour of mine – the two balls are correlated. (This happen when particles are split and their spins remain correlated: if I measure the spin of one particle, I automatically know that of the other.) Something similar often happens, and happened, in (mostly Leftist) politics. I am not sure which position I should take in a particular political struggle, but when I learn the position of my enemy, I automatically assume the opposite one. One should add that Lenin provided a scathing critique of this stance (ironically, his target was Rosa Luxemburg[6]). Such was the case in the cultural Cold War: when, in the late 1940s, Western culture was perceived as promoting universalist cosmopolitanism (under Jewish influence), pro-Soviet Communists from the USSR to France decided to turn patriotic, promoting their own cultural tradition and attacking imperialism for destroying it.

Is not something similar going on in the reaction to the Catalonia referendum that put Spain in turmoil in late 2017? Remember how Putin proclaimed the disintegration of the Soviet Union a mega-catastrophe – but now he supports Catalonian independence. The same holds for all those European Leftists who opposed the disintegration of Yugoslavia as the result of a dark German–Vatican plot – now, however (as with Scotland), separation is OK. And the Western centrist liberals are no better: always ready to support any separatist movement that threatens the geopolitical power of Russia, they now warn against the threat to the unity of Spain (hypocritically deploring the police violence against Catalonian voters, of course). In Slovenia, this confusion has reached its peak: the old Left, which to the very end was mostly against Slovene independence, pleading for a renewed, more open Yugoslavia, is now organizing petitions and demonstrations for Catalonia, while the nationalist Right, which fought for full Slovene independence, is now discreetly for the unity of Spain (since their conservative colleague Mariano Rajoy is the Spanish Prime Minister). Shame on the European establishment: obviously, some have the right to sovereignty and others not, depending on geopolitical interests.

One argument against Catalonian independence nonetheless seems rational: is Putin's support not obviously part of his strategy to strengthen Russia by way of working for the disintegration of European unity? Should, then, partisans of a strong, united Europe not advocate the unity of Spain? Here, one should dare to turn this argument around. Support for the unity of Spain is also part of the ongoing drive to assert the power of nation states against European unity. What we need in order to accommodate new local sovereignties (of Catalonia, of Scotland, maybe, and so on) is thus simply a stronger European Union: nation states should accustom themselves to more modest roles as intermediators between regional autonomies and a united Europe. In this way, Europe can avoid debilitating conflicts between states and emerge as a much stronger international agent, on a par with other big geopolitical blocs.

The failure of the EU to take a clear stance on the Catalonian referendum is just the last in the series of blunders, the biggest being the total lack of coherent political strategy towards the flow of refugees from the Middle East and north Africa into Europe. What's gone on lately in Yemen and in Syria (the systematic destruction of the whole of Yemen, the terrifying suffering of civilians in Ghouta, etc.) is a new example of how future immigrants are created – now is the time to do something, not waiting for the new immigrant wave when we will be able to resume our humanitarian battles. The confused reaction to the arrival of refugees failed to take into account the basic difference between immigrants and refugees: immigrants come to Europe to search for work, to meet the demand for a workforce in developed European countries, while refugees don't primarily come to work but simply to look for a safe place to survive – they often don't even like the new country in which they find themselves. Refugees who used to gather in Calais are paradigmatic here: they wanted not to stay in France but to move on to the UK. The same holds for the countries which most resist accepting refugees (the new 'axis of evil', Croatia, Slovenia, Hungary, the Czech Republic, Poland, the Baltic countries and Austria): they are definitely not the places where refugees want to settle. But perhaps the most absurd effect of this confusion is that Germany, the only country which behaved in a half-decent way towards refugees, became the target of many critics, not only Rightist defenders of Europe but also Leftists: in a typical super-ego way, they focused on the strongest link in the chain, attacking it for not being even stronger.

The most worrying aspect of the Catalonian crisis was thus the inability of Europe to take a clear stand: either to allow its member states to adopt their own politics with regard to separatism or refugees, or to take effective measures against those who don't want to accept collectively made decisions. Why is this so important? Europe is supposed to work as a minimal unity, supporting single states, providing a safety net for their tensions. Only such a Europe can be an important agent in the emerging New World Order, where the powerful agents are decreasingly single states. It is clearly in the interest of the US and Russia to weaken the EU or even to trigger its disintegration: a power vacuum will then be created which will be filled by new alliances of single European states with Russia or with the US. Who in Europe would like to see this?

The Eternal Return of the Same Class Struggle

The tension between global space and nation states lies in this defence of a specific (ethnic, religious, cultural) way of life, which is perceived as threatened by globalization; and the whole question of protecting one's way of life is problematic in all its versions, including the 'progressive' ones. Recall the polemics in the US concerning the statues of Robert E. Lee – was Lee a Southern gentleman who just fought for a certain way of life? A popular image of the Southern gentleman exists even in 'progressive' literature, from Horace in Lillian Hellman's *The Little Foxes*, a benevolent patriarch with a weak heart who is horrified by his wife's plans for the brutal capitalist exploitation of their property, to Atticus Finch in Harper Lee's *To Kill a Mockingbird*, who, as is revealed in the sequel, also had a dark racist underside. So all of a sudden Confederacy was not about slavery but about protecting a local 'way of life' from the brutal capitalist onslaught. These iconic Left-liberal figures of conservative bucolic-patriarchal anti-capitalism sincerely help Southern blacks when they are oppressed and falsely accused; however, their sympathy stops when blacks begin not only to fight but also to question the actual freedom provided by the Northern liberal establishment.

But Robert E. Lee was not even such a gentleman. There are no reports that he had any inner qualms about slavery. Furthermore, even among slave owners there was a division between those who, when they were reselling their slaves, took care that families with children remained together, and those who didn't bother about this and separated them – Lee belonged to this second, much harsher group. He may well have been a gentleman with nice manners and personal honesty, but he nonetheless dealt brutally with slaves – the difficult thing to accept is that the two characteristics go together.

A true white gentleman was executed on the order of Robert E. Lee: John Brown, one of the key political figures in the history of the US, the fervently Christian abolitionist who came closest to introducing radical emancipatory-egalitarian logic into the US political landscape. As Margaret Washington, a noted historican of the US, put it, he made it very clear that he saw no difference between whites and blacks, and 'he didn't make this clear by saying it, he made it clear by what he did'[7] – this is how a true gentleman talks and acts, if the term 'gentleman' can be given an emancipatory dimension. His consequential egalitarianism led him to get engaged in the armed struggle against slavery: in 1859, he tried to arm slaves and thus create a violent rebellion against the South; the revolt was suppressed and Brown was taken to jail by a federal force led by none other than Robert E. Lee. After being found guilty of murder, treason and inciting a slave insurrection, Brown was hanged on 2 December. Even today, long after slavery was abolished, Brown is a dividing figure in the American collective memory: his only statue, which stands on an obscure location in the Quindaro neighbourhood of Kansas City (the original town of Quindaro was a major stop on the Underground Railroad), was often vandalized.

So it goes without saying that all great American founding myths should be re-analysed: there is another, dark side to the War of Independence, to Alamo, and so on. The 'heroes of Alamo' were also defending slave ownership. This other side is portrayed in an interesting film of 1999, Lance Hool's *One Man's Hero*, which tells the story of Jon Riley (played by Tom Berenger) and the Saint Patrick's Battalion, a group of Irish Catholic immigrants who deserted from the mostly Protestant US Army to the Catholic Mexican side during the Mexican-American War of 1846–8 and fought heroically to defend the Republic of Mexico from US aggression. At the movie's end, while working in a stone quarry for military prisoners, Riley is told by his former US commander that he has been freed, to which he responds, 'I have always been free.'

The point is not just to debunk the War of Independence as fake: there undoubtedly is an emancipatory dimension in the works of Jefferson, Paine, and so on. In spite of being a slave owner, Jefferson is an important link in the chain of modern emancipatory struggles, and one is justified in claiming that the struggle for the abolition of slavery was basically the continuation of Jefferson's work. Jefferson was a different kind of man from Robert E. Lee, and the inconsistencies in his position just demonstrate how the American revolution is an unfinished project (as Habermas would have put it). In some sense, its true conclusion, its second act, was the Civil War; in another sense, it was over only in 1960, with the realization of the black right to vote; and in another sense, as the persistence of the Confederacy myth demonstrates, it is not yet over today. (Similarly, although Immanuel Kant's views are racist, he nonetheless contributed to the process which led to contemporary emancipatory struggles – to put it bluntly, there is no Marxism and no socialism without Kant.) This is the point missed by Trump when he placed 'respect' for Lee within the canon of respect for American tradition and asked where all this will stop – first Lee, then Washington, then . . . What lurks beneath the fight for statues of Lee is simply the refusal to bring the American revolution to an end.

But there is another aspect to Trump's proclamations which is as a rule ignored: his reluctance unambiguously to condemn alt-right violence and his repeated claims that 'Both sides are guilty' strangely mirror the Leftist multiculturalist strategy ('True, ISIS is committing horrible crimes – but do we not do similar evil things? Who are we to judge them?'). As Jamil Khader pointed out in a crucial intervention,[8] in his reactions to the Charlottesville killing Trump displayed not only multiculturalism but also, and above all, the emancipatory legacy of universalism.[9] This point was also missing in most liberal and Leftist responses to Trump's comments on slavery and white supremacy – that:

> no identity can easily fill in the empty space of universality with its proper content, and that identities should always be taken up to fulfill the promises of the immanent universal dimension that exists in the form of a gap at their core. Radical, if not revolutionary change, can happen only when liberals and leftists rethink their conception of identity in light of this repressed universal dimension at its core . . . The problem is that mainstream liberal and leftist discourses on identity politics and political correctness have shifted the struggle for justice, freedom, and equality from oppression and exploitation to tolerance and respect under the banner of a post-racial ideology . . . Trump's other controversial statements about moral equivalency between Neo-Nazi White supremacist terrorists and antifa activists did not emerge out of a vacuum. Indeed, his points about violence on 'many sides' and that there were 'some very fine *people* on both *sides*' are symptomatic of the same humanist strategies that liberals and leftists had used during the culture and canon wars to relativize conflicts, subjectivize the Other (giving the evil Other a voice and a human story), and remain on neutral grounds.[10]

Along the same lines, Walter Benn Michaels wrote apropos the (often ridiculous) polemics about cultural appropriation:

> even our own stories don't belong to us – no stories belong to anyone. Rather, we're all in the position of historians, trying to figure out what actually happened . . . Identity crimes – both the phantasmatic ones, like cultural theft, and the real ones, like racism and sexism – are perfect for this purpose, since, unlike the downward redistribution of wealth, opposing them leaves the class structure intact . . . The problem is not that rich people can't feel poor people's pain; you don't have to be the victim of inequality to want to eliminate inequality. And the problem is not that the story of the poor doesn't belong to the rich; the relevant question about our stories is not whether they reveal someone's privilege but whether they're true. The problem is that the whole idea of cultural identity is incoherent, and that the dramas of appropriation it makes possible provide an increasingly economically stratified society with a model of social justice that addresses everything except that economic stratification.[11]

Benn Michaels is fully justified here: of course we should fight White liberal cultural appropriations, but not simply because they practise imbalance in the cultural exchange – we should fight them because they practise the struggle for emancipation in such a way that they ignore and neutralize its key dimension. And the same holds for the feminist struggle. In the last decades, a new form of feminism has risen to prominence, especially in the US, which one cannot but designate as 'neo-liberal feminism';[12] its three main features are: (1) individualization of persistent gender inequality (today, gender inequality is not systemic but mostly a consequence of individual choices, so there is no need for structural analysis and large social changes); (2) privatization of political responses (solutions must be individual); (3) liberation through capitalism (women can achieve and ensure gender equality through the free market: 'the feminist is the entrepreneur, capable of competing alongside of men, and winning or losing in the marketplace.'[13] The appeal of this approach resides also in the pleasures it promises: those of avoiding conflict (organized political struggle), of indulging in consumption and financial success, and so on.[14] Do we not have here an exemplary case of hegemonic rearticulation in which feminism is included in a different chain of equivalences? If this process of rearticulation is open and ultimately contingent, we cannot claim that neo-liberal feminism is a 'betrayal' of the 'true' feminism which links feminine liberation to the universal emancipation of all those who are exploited. Is, then, this feminism a concrete universality which transforms itself into new figures, where we should not introduce a critical distinction between radical and bourgeois feminisms but see different feminisms as particular moments, each of which contributes new content, opens up new spaces of political practice, and simultaneously implies specific limitations? If not, why, exactly, not? Because class struggle is the only universal antagonism, an antagonism that cuts across the entire social edifice, the impossible/real which casts its shadow on all other antagonisms.

The basic premise of classical Marxism (the premise that grounds its call for the 'unity of theory and practice') is that, by its objective social position (that of the 'part of no-part'(Rancière) of the social edifice, the point of its 'symptomal torsion'(Badiou)), the working class is pushed towards a correct insight into the state of society (its basic antagonisms) and, simultaneously, towards the action to be taken to set it straight (the revolutionary transformation). Does this still hold today? Does the rise of the populist fury and rage not bear witness to an irreducible break in the 'unity of theory and practice'? It is as if the 'objective' social position of those exploited and marginalized no longer pushes them towards a clear 'cognitive mapping' of their predicament, which would engage them in a universal emancipatory struggle, but rather expresses itself in frustrated and occasionally violent impotence, betraying their loss of basic orientation. So, instead of a united front, local lower classes fear immigrants, who take refuge in fundamentalism, while the trade unions fight for the welfare of those whom they represent more often against other parts of the working class than against capital – can one imagine here a united front? The projected unity is necessarily and continuously undermined by the counter-force that is immanent to the ongoing process of class struggle: the conflict between local lower classes and immigrants (or between feminist struggle and workers struggle) is not an externally imposed abomination caused by the manipulations of enemy propaganda, but the form of appearance of the same class struggle. Local workers perceive immigrants as the stooges of big capital, brought into the country to undermine their strength and to compete with them since their wages are lower; immigrants see local workers, even if they are poor, as part and parcel of the Western order that marginalizes them. No easy preaching about how they are actually on the same side can be effective in such a situation where competition is real.

Therein resides the fatal limitation of the attempts to counter the rise of Rightist populism with Leftist populism, a populism that would listen to the real concerns of ordinary people instead of trying to impose on them some high theoretical vision of their historical task. The fears, hopes, and problems 'real people' experience in their 'real lives' always appear to them as moments of a certain ideological vision, i.e. as Althusser saw it well, ideology is not a conceptual frame externally imposed on the wealth of reality, it is our experience of reality itself. To break out of ideology, it is not enough to get rid of the distorting ideological lenses – hard theoretical work is needed.

To get a taste of the complexity of this struggle, let's take a recent example of the conflict between different emancipatory demands. On a US campus, an incident took place recently:[15] a group of young Latino workers were restoring the façade of a house on a plateau that overlooked a nearby swimming pool where a group of young middle-class women were sunbathing in bikinis, and the workers started to direct flirtatious comments at them (what in Latin America they call *el piropo*). Predictably, the women felt harassed, and they complained, and the solution imposed by the authorities was no less predictable: they separated the house from the pool area by a plastic wall, and they constructed a special plastic tunnel through which the workers had to approach their workplace, cutting off the view of the pool area – a perfect example of the politically correct way of dealing with sexism which just reinforces the lines separating groups of people.

From the women's standpoint, what happened was a clear-cut case of male-chauvinist harassment 'objectivizing' women as sexual prey, while from the workers' standpoint their exclusion was a no less clear-cut example of maintaining a class distinction, of protecting the white middle class from contact with ordinary workers. Is it then a case of feminist struggle versus class struggle, with the long-term solution being to somehow unite the two and convince both sides that their respective struggles are moments of the same universal struggle for emancipation? It's not as simple as that, since it is the class struggle itself that overdetermines the tension between the two struggles: the workers' *piropo* was obviously so disturbing to the girls because it came from lower-class boys unworthy of their attention, and the boys were aware of this dimension when they were reprimanded. Feminism can also play a class game, implying that lower classes are vulgar, male-chauvinist, not politically correct, so that a fear of being 'harassed' reveals itself to be the fear of lower-class vulgarity. This, however, in no way means that we should say to the women, 'Endure the harassment on account of solidarity with the working class (and remember they are Latino foreigners who have their way of life)!' – at this level, in the direct confrontation of the two views, the conflict cannot be resolved, and *this irresolvable deadlock IS the reality of class struggle.*

Recognizing the overdetermining role of class struggle does not amount to accepting the standard 'essentialist'-Marxist claim that sexuality gets violent owing to class struggle but remains in itself non-violent – class struggle co-opts the immanent violence and deadlocks that pertain to sexuality as such. In the same way, other particular struggles obey their own immanent antagonist logic: for example, different ethnic-religious 'ways of life' are immanently out-of-sync due to the different mode of regulating collective *jouissance*, while human industry affects our environment in potentially dangerous ways independently of specific modes of production. Class struggle does not introduce the antagonism but overdetermines the immanent antagonism. More precisely, class antagonism is doubly inscribed – it encounters itself in its oppositional determination, among the struggles whose totality it overdetermines. Back to our example, class struggle is represented by the resistance towards Mexican workers by the bathing girls (in contrast to their feminist claims), plus it overdetermines the very articulation of these particular struggles. The actuality of the class struggle is the tension between the two emancipatory struggles – but, again, not in the sense that the workers stand for the proletariat and the girls for the bourgeoisie. If one had to decide to which side one should give priority in the conflict, there are strong arguments that the bathing girls effectively were harassed and should be somehow protected. The overall dynamic of class struggle is the overdetermining factor of the conflict and, consequently, that which makes the conflict irresolvable in its own terms (even if we give the priority to the harassed girls, there is a shadow of injustice in this choice). And the same goes for the opposite choice: class struggle is that which makes also the 'class' choice of Mexican workers over bathing women unjust. Paradoxically, class struggle is itself the factor that limits the scope of direct reference to class struggle . . .

The formal feature that makes class struggle exceptional is that it cannot be reduced to a case of identity politics: while the goal of feminism is not to destroy men but to establish new, more just rules as to how the two sexes should interact, and while the most aggressive religious fundamentalism wants to assert itself by way of destroying other religions, proletarian class struggle aims at abolishing class *difference*, eliminating not only the ruling class but also itself – the aim of proletarian struggle is to create conditions in which proletarians themselves would cease to exist. (Along the same lines, John Summers has pointed out how multiculturalism emerged as the ideology of corporate elites: a politics directed at gender, race or any other identity is a game lost in advance. The struggle for identity is a perfect substitute for the class struggle, since it keeps people in permanent mutual conflict, while the elite withdraws and observes the game from a safe distance.[16]) A recent analysis published in the *Guardian* brings out the basic inconsistency of identity politics:

> Many on the left had become acutely aware that colour blindness was being used by conservatives to oppose policies intended to redress historical wrongs and persisting racial inequities. With the collapse of the Soviet Union, the anti-capitalist economic preoccupations of the old Left began to take a backseat to a new way of understanding oppression: the politics of redistribution was replaced by a 'politics of recognition'. Modern identity politics was born. As Oberlin professor Sonia Kruks writes, 'What makes identity politics a significant departure from earlier [movements] is its demand for recognition on the basis of the very grounds on which recognition has previously been denied: it is qua women, qua blacks, qua lesbians that groups demand recognition . . . The demand is not for inclusion within the fold of

'universal humankind' . . . nor is it for respect 'in spite of' one's differences. Rather, what is demanded is respect for oneself as different.'

When liberal icon Bernie Sanders told supporters, 'It's not good enough for somebody to say, "Hey, I'm a Latina, vote for me,"', Quentin James, a leader of Hillary Clinton's outreach efforts to people of colour, retorted that Sanders's 'comments regarding identity politics suggest he may be a white supremacist, too'. This brings us to the most striking feature of today's right-wing political tribalism: the white identity politics that has mobilized around the idea of whites as an endangered, discriminated-against group. People want to see their own tribe as exceptional, as something to be deeply proud of; that's what the tribal instinct is all about. For decades now, non-whites in the United States have been encouraged to indulge their tribal instincts in just this way, but, at least publicly, American whites have not.[17]

Identity politics reaches its peak (or, rather, its lowest point) when it refers to the unique experience of a particular group identity as the ultimate fact that cannot be dissolved in any universality: 'only a woman/lesbian/trans/Black/Chinese knows what is it to be a woman/lesbian/trans/Black/Chinese.' While this is true in a certain trivial sense, one should thoroughly deny any political relevance to it and shamelessly stick to the old Enlightenment axiom: all cultures and identities can be understood – one just has to make an effort to get it.[18] The secret of identity politics is that, in it, a white/male/hetero position remains a universal standard, everyone understands it and knows what it means, which is why it is the blind spot of identity politics, the one identity that it is prohibited to assert. Sooner or later, however, we get the return of the repressed: the white/male/hetero identity breaks out and begins to play the same card – 'Nobody really understands us, you have to be a white/hetero/male to understand what it means to be a white/hetero/male . . . ' What these reversals prove is that one cannot get rid of universality so easily. The obvious old Marxist point about how there is no neutral universality, i.e. about how every universality that presents itself as neutral obfuscates and thereby privileges actual privileges, should not seduce us into abandoning universality as such – if we do this, we obliterate the fact that our very argumentation against false universalities speaks from the position of true universality (which enables us to perceive the position of the underprivileged as unjust). Paradoxically, the assertion of white/hetero/male identity would deprive them of their implied universality and compel them to accept their particularity.

Such an assertion may appear to play directly into the hands of white supremacists – but does it? Everyone who is troubled by the new anti-immigrant populism should make the effort to watch *Europa: The Last Battle* (Tobias Bratt, Sweden, 2017), a ten-part documentary that can easily be downloaded for free. It presents *in extenso* the neo-Nazi version of the last hundred years of European history: it was dominated by Jewish bankers who control our entire financial system; from the beginning, Judaism stood behind Communism, and the wealthy Jews directly financed the October Revolution to deal a mortal blow to Russia, a staunch defender of Christianity; Hitler was a peaceful German patriot who, after being democratically elected, changed Germany from devastated land to a welfare country with the highest living standard in the world by withdrawing from the international banking controlled by Jews; international Jewry declared war on him, although Hitler desperately strived for peace; after the failure of the European Communist revolutions in the 1920s, the Communist centre realized that one has first to destroy the moral foundations of the West (religion, ethnic identity, family values), so it founded the Frankfurt School, whose aim was to denounce family and authority as pathological tools of domination, and to undermine every ethnic identity as oppressive. Today, in the guise of different forms of Cultural Marxism, their efforts are finally showing results; our societies are caught in eternal guilt for their alleged sins, they are open to unbridled invasion of immigrants, lost in empty hedonist individualism and lack of patriotism. This corruption is secretly controlled by Jews like Soros, and only a new figure like Hitler, who would re-awaken our patriotic pride, can save us . . . When one watches this spectacle, one cannot avoid the impression that, although the authors went much further than our average racist populists would be ready to go, we are getting in *Europa* a kind of 'absent centre' of the multitude of communitarian-populist movements that currently thrive, the zero-point towards which they all tend and in which they would converge.

When, in my critique of this tendency, I claimed that the greatest threat to Europe are its populist/racist defenders, I was reproached for the obvious absurdity of this claim: how can those who want to defend Europe pose a threat to it? In principle, the answer is easy to give: the Europe these defenders try to save (a neo-tribal Europe of fixed ethnic identities) is the negation of all that is great in the European legacy. (The obvious anti-Eurocentric reproach to my claim is, of course, that Europe, the agent of global colonial domination, has no right to offer its ideological foundations as a possible weapon against racism.) There is some truth in this – no wonder that the most radical 'defenders' of Europe look with distrust at Christianity and prefer pagan (Celtic, Nordic) spirituality. And one can easily see where the problem resides – even those who still pay lip service to Christian Europe advocate a weird Christianity with a distinct pagan twist. As recently reported, Viktor Orbán has declared

> the end of 'liberal democracy' in Hungary, saying it has failed to defend freedoms and Christian culture in the wake of the migrant crisis. He vowed to build a 'Christian democracy' defying EU dictates. 'The era of liberal democracy has come to an end. It is unsuitable to protect human dignity, inadmissible to give freedom, cannot guarantee physical security, and can no longer maintain Christian culture,' Orbán said.[19]

Are these statements not difficult to combine with statements like the following from *Galatians* 3:28: 'There is neither Jew nor Gentile, neither slave nor free, nor is there male and female, for you are all one in Christ Jesus.' And how would Christian defenders of the family deal with the famous passage from *Matthew* 12:46–50:

> While he was still talking to the multitudes, behold, his mother and brothers stood outside, seeking to speak with him. Then one said to him, 'Look, your mother and your brothers are standing outside, seeking to speak with you.' But he answered and said to the one who told him, 'Who is my mother and who are my brothers?' And he stretched out his hand toward his disciples and said, 'Here are my mother and my brothers! For whoever does the will of my father in heaven is my brother and sister and mother.'

There is, however, another higher-level counter-argument often evoked against immigrants: the point is not that, in their way of life, they are different from us, but that *they* have problems with difference (the coexistence of different ways of life) as such. The exemplary case is here that of the Dutch Rightist populist politician Pim Fortuyn, killed in early May 2002, two weeks before elections in which he was expected to gain one fifth of the votes: a Rightist populist whose personal features and even (most of his) opinions were almost perfectly politically correct – he was gay, had good personal relations with many immigrants, and had an innate sense of irony . . . in short, he was a good tolerant liberal with regard to everything *except his basic political stance*: he opposed fundamentalist immigrants because of their hatred towards homosexuality, women's rights, etc.

The reply is, of course, that this argument relies on meta-racism, i.e. on a more subtle form of racism in which we assert our superiority over the Other precisely by claiming that our Other, not us, is racist . . . But there is another more basic problem we are dealing with here: asserting openness and fluidity of identities is not enough, and it is indeterminacy which is pushing people towards proponents of populist ethnic identity. The tough question is therefore: what kind of identity is acceptable for a radical Leftist? Abstract universalism doesn't work, as was made clear by, among others, Claude Lévi-Strauss, who, in the essays collected in the second volume of his *Structural Anthropology*,[20] forcefully demonstrated how a strong assertion of one's ethnic identity and even of its superiority to others does not necessarily imply racism. He shows that many tribes who call themselves 'human' (with regard to other tribes to whom one denies this quality), i.e. in whose language the word for 'human' is the same as the word for 'belonging to our tribe', are not racist in the modern sense of the term. Although they may appear offensively racist, upon a closer look their stance is much more modest: it should be read as an implicit assertion of being caught into one's own way of life – 'We are what we are, and for us this is what being human means; we cannot step out of our world to judge us and others from nowhere, so we also let others be.' In short, their assertion of self-identity is not negatively mediated by others in the form of envy.

In order to mask its own divisions, populist identity is based on the negative reference to the Other: no Nazi without a Jew, no European without the immigrant threat, etc. However, political correctness is also grounded in a negative reference, parasitizing on the sexist/racist 'incorrect' Other – this is why the politically correct subjectivity is a mixture of eternal self-guilt (searching for the remainders of sexism or racism in oneself) and arrogance (constantly reprimanding and judging the guilty others). The paradox is thus that the problem of populist fundamentalism does not reside in the fact that it is too identitarian (against which we should emphasize fluidity and contingency of every identity) but, on the contrary, in the fact that it lacks proper identity, that its identity clings on to a denial of its constitutive Other. Are the so-called fundamentalists, be it Christian or Muslim, really fundamentalists in the authentic sense of the term? Do they really believe? What they lack is a feature that is easy to discern in all authentic fundamentalists, from Tibetan Buddhists to the Amish in the US: the absence of resentment and envy, the deep indifference towards the non-believers' way of life. If today's so-called fundamentalists really believe they have found their way to Truth, why should they feel threatened by non-believers, why should they envy them? When a Buddhist encounters a Western hedonist, he hardly condemns him or her – he just benevolently notes that the hedonist's search for happiness is self-defeating. In contrast to true fundamentalists, the pseudo-fundamentalists are deeply bothered, intrigued, fascinated, by the sinful life of the non-believers. One can feel that, in fighting the sinful other, they are fighting their own temptation. This is why the so-called Christian or Muslim fundamentalists are a disgrace to true fundamentalism.

Does this mean that we should simply tolerate a peaceful coexistence of different ways of life? Unfortunately, this is no solution. We should persist in the properly dialectical approach: such an acceptance of identity in no way invalidates universality, it merely renders it 'concrete' in the Hegelian sense. When white supremacists say, 'We only want for us what the supposedly marginalized others demand for themselves – to freely assert and develop our identity, our way of life,' there is nothing wrong with this statement as such. The problem is that they do not mean just that but much more, implicitly privileging their own way of life at the expense of others – in short, the problem is in their implicit universality. Each way of life implies its own universality: it is not just about itself but also about how to relate to others, and the two cannot be separated. Western liberal multiculturalism is different, say, from co-existence of religions and ethnic groups in India; the problem (not only) with Islam is how it relates to other religions and cultures (and atheism) in its own countries – are they tolerated as equal, can they act in the public space? When Western liberals prohibit certain sexual practices (not only) of Muslims, like arranged marriages against the will of the woman involved, does the state have the right to intervene, or is this an intrusion into another's way of life? The problem is that the relationship between different ways of life is always also a conflict of universalities – there is no universal neutral space exempted from it.

The only true emancipatory gesture is therefore to persist in the search for universality (as, for example, Malcolm X did). And the white person should cast a self-critical glance on their position, of course, but without getting caught in the vicious cycle of eternal guilt. The prohibition of asserting the particular identity of White Men (as the model of oppression of others), although it presents itself as the admission of their guilt, confers on them a central position: this very denial of the right to assert their particular identity makes them into the universal-neutral medium, the place from which the truth about the others' oppression is accessible. And this is why white liberals so gladly indulge in self-flagellation: the true aim of their activity is not really to help the others but the *Lustgewinn* brought about by their self-accusations, the feeling of their own moral superiority over others. The problem with the self-denial of white identity is not that it goes too far but that it does not go far enough: while its enunciated content seems radical, its position of enunciation remains that of a privileged universality.

When we try to clarify how to relate the universal struggle for emancipation to the plurality of ways of life, nothing should be left to chance, not even the most self-evident general notions. Left-liberals view the very notion of 'way of life' with suspicion (if it doesn't relate to marginal minorities, of course), as if it conceals a proto-fascist poison; against this suspicion, one should accept the term in its

Lacanian version, as something that points beyond all cultural features towards a core of the Real, of *jouissance* – a 'way of life' is ultimately the way in which a certain community organizes its *jouissance*. This is why 'integration' is such a sensitive issue: when a group is under pressure to 'integrate' into a wider community, it often resists out of fear that it will lose its mode of *jouissance*. A way of life does not merely encompass rituals of food, music and dance, social life, and so on, but also, and above all, habits and written and unwritten rules of sexual life (inclusive of rules of mating and marriage) and of social hierarchy (respect for elders, and so on). In India, for example, some post-colonial theorists even defend the caste system as part of a specific way of life that should be protected from the global onslaught of global individualism.

To solve this problem, the preferred vision is that of a united world with all its particular ways of life thriving, each of them asserting its difference from others, not as an antagonistic relationship, not at the expense of others, but as a positive display of creativity that contributes to the wealth of the whole of society. When an ethnic group is prevented from expressing its identity in this creative way because it is under pressure to renounce it and 'integrate' itself into the predominant (usually Western) culture and way of life, it cannot but react by withdrawing into negative difference, a regressive, purist fundamentalism that fights the predominant culture, including by violent means – in short, fundamentalist violence is a reaction for which the predominant culture is responsible.

This entire vision of creative differences, of particular identities contributing to a united world, threatened by the violent pressure on the minorities to 'integrate' – in other words, by the false universality of the Western way of life, which imposes itself as a standard for all – is to be rejected in its entirety. The world we live in is one, but it is such because it is traversed (and, in a way, even held together) by the same antagonism that is inscribed into the very heart of global capitalism. Universality is not located over and above particular identities, it is an antagonism that cuts from within each 'way of life'. This antagonism determines all emancipatory struggles: explicit and unwritten rules of hierarchy, homophobia, male domination and so on are key constituents of the 'way of life' in which such struggles occur. Let's take the very sensitive case of China and Tibet: the brutal Chinese colonization of Tibet is a fact, but this fact should not blind us to what kind of country Tibet was before 1949, and even before 1959 – a harsh feudal society with an extreme hierarchy regulated in detail. In the late 1950s, when the Chinese authorities still more or less tolerated the Tibetan 'way of life', a villager visited his relatives in a neighbouring village without asking his feudal master for permission. When he was caught and threatened with severe punishment, he took refuge in a nearby Chinese military garrison, but when his master learned of this he complained that the Chinese were brutally meddling in the Tibetan way of life – and he was right! So what should the Chinese do? Another similar example is that of a traditional Tibetan custom: when a serf met a landowner or a priest on a narrow path, he

> would stand to the side, at a distance, putting a sleeve over his shoulder, bowing down and sticking out his tongue – a courtesy paid by those of lower status to their superiors – and would only dare to resume his journey after the former serfs had passed by.[21]

In order to dispel any illusions about Tibetan society, it is not enough to note the distasteful nature of this custom. Over and above the usual stepping-aside and bowing, the subordinated individual – to add insult to injury, as it were – had to fix his face in an expression of humiliating stupidity (open-mouthed, tongue sticking out, eyes turned upwards) in order to signal with this grotesque grimace his worthless stupidity. The crucial point here is to recognize the violence of this practice, a violence that no consideration of cultural differences and no respect for otherness should wash over. Again, in cases like these, where does the respect for the other's way of life reach its limit? True, we should not intervene from outside, imposing our standards, but is it not the duty of every fighter for emancipation unconditionally to support those in other cultures who, from within, resist such oppressive customs?[22]

Anti-colonialists as a rule emphasize how the colonizers try to impose their own culture as universal and thereby undermine the indigenous way of life; but what about the opposite strategy, which resides in strengthening local traditions in order to make colonial domination more efficient? No wonder the British colonial administration of India elevated *The Laws of Manu* – an ancient detailed justification for and manual of the caste system – into the seminal text to be used as a reference for establishing the legal code that would render possible the most efficient domination of India; up to a point, one can even say that *The Laws of Manu* only became *the* book of the Hindu tradition retroactively. And, in a more subtle way, the Israeli authorities are doing the same on the West Bank: they silently tolerate (or at least do not investigate seriously) 'honour killings', being well aware that the true threat to them comes not from devout Muslim traditionalists but from modern Palestinians.

This is the lesson that not only refugees but all members of traditional communities should learn: the way to strike back at cultural neo-colonialism is not to resist it on behalf of their traditional culture but to reinvent a more radical modernity – something Malcolm X, again, was well aware of. It is this unreadiness to accept the primary role of universality which saps the majority of post-colonial studies. Ramesh Srinivasan's work[23] is representative of the effort to 'decolonize' digital technology, which is not just a neutral-universal technological framework for the exchange between cultures: it privileges a certain (Western modern) culture, so that even benevolent efforts to extend computer literacy and include everyone in the digital 'global village' secretly prolong colonization, insisting on the integration of the subaltern into Western modernity and thus oppressing their cultural specificity. Srinivasan mentions briefly that communities themselves are 'multifaceted and diverse', but instead of developing this point into the notion of antagonisms that traverse every community, he waters it down into global relativization and the partiality of every view. The basic units of his vision of reality are communities which, through their life practices, form their own vision of reality; they are the starting point, and 'conversations that surpass the bounds of community' come second, so that when we practise them we should always be careful that we respect the authentic voice of the particular community. Therein lies the trap of the popular notion of the 'global village': it imposes on particular non-Western communities assumptions which are not theirs, that is, it practices cultural colonialism:

> While it is important to learn about other people, cultures, and communities on their terms, we must respect the power and importance of local, cultural, indigenous, and community-based creative uses of technology. Conversations that surpass the bounds of community can and should emerge but only when the voices of their participants are truly respected. From this perspective, the 'global village' is the problem rather than the solution. We must reject assumptions about technology and culture that are dictated by Western concepts of cosmopolitanism.[24]

This is why Srinivasan criticizes Ethan Zuckerman, who

> is correct to say that many of today's challenges, such as climate change, require global conversation and cross-cultural awareness. But not all challenges are global and indeed thinking globally about people's traditions, knowledges, struggles and identities may unintentionally exclude them from positions of control and power.[25]

So, again, the global view is strictly secondary; what comes first is the multiplicity of local communities with their particular 'ontologies'. And even modern science in its global reach is historically relativized as one among many fields of knowledge with no right to be privileged – Srinivasan approvingly quotes Boaventura de Sousa Santos, who claims:

> the epistemological privilege granted to modern science from the seventeenth century onwards, which made possible the technological revolutions that consolidated Western supremacy, was also instrumental in suppressing other non-scientific forms and knowledges ... It is now time to build a more democratic and just society and ... decolonize knowledge and power.[26]

It would be easy to show that such 'fluid ontology' of the multiplicity of cultures is grounded in a typically Western postmodern view based on the historicization of all knowledge, a view that has nothing to do with actual premodern societies. But much more important is the link between Srinivasan's disavowal of universality (his insistence on the primacy of particular cultures/communities) and his ignoring of the inner antagonisms constitutive of particular communities: they are the two sides of the same misrecognition, since universality is not a neutral entity elevated above particular cultures; it is inscribed into them, at work in them, in the guise of their inner antagonisms, inconsistencies and disruptive negativities. Every particular way of life is a politico-ideological formation whose task is to obfuscate an underlying antagonism, a particular way to cope with this antagonism, and this antagonism traverses the entire social space. Apart from some tribes in the Amazon jungle who have not yet established contact with modern society, all communities today are part of a global civilization in the sense that their autonomy itself has to be accounted for in terms of global capitalism. Let's take the case of the native American tribes' attempts to resuscitate their ancient way of life. This way of life was derailed and thwarted by their contact with modern civilization, which had the devastating effect of leaving the tribes totally disoriented, deprived of a stable communal framework; and their attempts to regain some stability in restoring the core of their traditional way of life as a rule depend on their success in finding their niche in the global market economy. Many tribes wisely spend the income earned from casinos and mining rights on this restoration or, as Richard Wagner put it, 'die Wunde schliesst der Speer nur der sie Schlug' (only the spear that struck you heals the wound).

In Orwell's *1984*, there is a famous exchange between Winston and O'Brien, his interrogator. Winston asks him:

> 'Does Big Brother exist?'
>
> 'Of course he exists. The Party exists. Big Brother is the embodiment of the Party.'
>
> 'Does he exist in the same way as I exist?'
>
> 'You do not exist,' said O'Brien.

Should we not say something similar about the existence of universality? As discussed earlier, to the nominalist claim that there is no pure neutral universality, we can say, 'No, today it's the particular ways of life that do not exist as autonomous modes of historical existence, the only actual reality is that of the universal capitalist system.' So, in contrast to identity politics, which focuses on how each group should be able to fully assert its particular identity, the radical task is to enable each group to have full access to universality – which does not mean recognizing that one is also part of the universal human genus, or asserting some ideological values which are considered universal. It means recognizing how one's own universality is at work in the fractures of one's particular identity, as the 'work of the negative' that undermines every particular identity – or, as Susan Buck-Morss put it, 'universal humanity is visible at the edges':[27]

> rather than giving multiple, distinct cultures equal due, whereby people are recognized as part of humanity indirectly through the mediation of collective cultural identities, human universality emerges in the historical event at the point of rupture. It is in the discontinuities of history that people whose culture has been strained to the breaking point give expression to a humanity that goes beyond cultural limits. And it is our emphatic identification with this raw, free, and vulnerable state, that we have a chance of understanding what they say. Common humanity exists in spite of culture and its differences. A person's nonidentity with the collective allows for subterranean solidarities that have a chance of appealing to universal, moral sentiment, the source today of enthusiasm and hope.[28]

Here Buck-Morss provides a precise argument against the postmodern poetry of diversity: it masks the underlying *sameness* of the brutal violence enacted by culturally diverse cultures and regimes: 'Can we rest satisfied with the call for acknowledging "multiple modernities", with a politics of "diversity", or "multiversality", when in fact the inhumanities of these multiplicities are often strikingly the same?'[29]

Furthermore, when Leftist liberals endlessly vary the motif of how the rise of terrorism is the result of Western colonial and military interventions in the Middle East, meaning that we are ultimately responsible for it, their analysis, although affecting respect towards others, stands out as a blatant case of patronizing racism that reduces the Other to a passive victim and deprives it of any agenda. What such a view fails to see is how Arabs are in no way just passive victims of European and American neo-colonial machinations. Their different courses of action are not just reactive, they are different forms of active engagement in their predicament: the expansive and aggressive push towards Islamization (financing mosques in foreign countries, for example), and open warfare against the West are ways of actively engaging in a situation with a well-defined goal.

For the same reason, one should also doubt the emancipatory value of referring to the people who were colonized as 'natives' or 'first people'. When, in the US, a hypothesis was ventured that today's native Americans ('Indians') were not the first human inhabitants there, that they displaced another earlier race, the predominant Left-liberal reaction was that this is a dark move to obfuscate the horrors of colonization ('what we, white people, did to the Indians, they did to others'). Similarly, anti-racists view with suspicion the historians trying to demonstrate that the first (Boer) white settlers in South Africa were there simultaneously (or even a couple of decades earlier) than today's black majority, which invaded the country from the north, displacing the original inhabitants (Bushmen and Hottentots). While these suspicions are justified, i.e., while the white racist stakes of such research are obvious, one should nonetheless absolutely reject the idea that proving that today's 'native Americans' or the black majority in South Africa were not the true 'first people' there in any way diminishes or undermines the Black or 'native American' anti-racist struggle for full emancipation. Today's racism has nothing to do with the historical question of 'who got there first?': it is a matter of today's relations of domination and exploitation.

The Western legacy is effectively not just that of colonial and post-colonial imperialist domination, but also that of self-critical examination of the violence and exploitation that the West brought to the Third World. The French colonized Haiti, but the French Revolution also provided the ideological foundation for the rebellion which liberated the slaves and established independent Haiti; the process of decolonization was set in motion when the colonized nations demanded for themselves the same rights that the West took for itself. In short, one should never forget that the West provided the very standards by which it (as well as its critics) measures its criminal past.

4

Ernst Lubitsch, Sex and Indirectness

Theodor Adorno turned around Benedetto Croce's patronizing historicist question, 'What is living and what is dead in the philosophy of Hegel?' (the title of his main work): if Hegel's thinking is still alive, then the question we should be asking now is not, 'How does Hegel's work stand with regard to our current constellation? How are we to read it so that it still says something to us?', but, 'How do *we today* stand with regard to – in the eyes of – Hegel?' Exactly the same holds for the film director Ernst Lubitsch: the question is, 'How would our contemporaneity appear in the eyes of Lubitsch?' Therein resides the actuality of Lubitsch: while, of course, rejecting with disgust populist neo-racism, he would immediately have perceived the falsity of its opposite, politically correct moralism, clearly seeing their hidden complicity. Lubitsch would have been appalled to witness how the perverse pleasures of obscenities, irony even, have moved to the Right, while the Left is increasingly caught up in pathetic, ascetic, puritanical moralism.

What this means is that there will be no renewal of the Left without a Lubitsch touch.

From Indirectness to Ratatatata

So how would Lubitsch counteract this unholy coupling? Through comic indirectness – but does this work? After the full extent of the Nazi atrocities became known to the public, Lubitsch's masterpiece *To Be or Not to Be* (1942), as well as Chaplin's *The Great Dictator*

(1940), were both criticized for downplaying the horrors of Nazism by making comedy out of it – Chaplin himself said that if he had been aware of the horror of the concentration camps, he would never have shot his film. However, the situation is much more complex and ambiguous. Isn't it that, in a tragedy, the victims retain a minimum of dignity; which is why, when horror crosses a certain line, to portray it as tragedy is to blasphemously downplay its extent? In Auschwitz (or in a gulag camp), victims were deprived of their human dignity to such a degree that they could no longer be perceived as tragic heroes; instead, a comic element came into play – no wonder that some of the best films about concentration camps are comedies. Should we, then, really be surprised that one of the jokes going around Sarajevo, when the city was besieged by Serbian forces from 1992 to 1995 and its gas supply was often cut off, was: 'What's the difference between Auschwitz and Sarajevo? In Auschwitz, they at least never ran out of gas.' Or what about the cruel quip popular among the survivors of the Srebrenica massacre of 1995, when more than 7,000 Bosnian men and boys were killed by Serbian forces? (To understand this joke, one has to remember that, in old Yugoslavia, when one went to a butcher to buy beef the butcher usually asked, 'With bones or without?' – bones were added to make the beef soup taste better.) After the war, a refugee returns from Germany to Srebrenica and wants to buy a piece of land there to build a house, so he asks a friend about the price, and the friend answers: 'It depends – do you want it with or without bones?' This is how you deal with trauma which cannot yet be properly mourned and come to terms with – you turn it into a joke. There is nothing disrespectful in this: on the contrary, such jokes imply an awareness that the memory is still too raw to apply to it the process of mourning.

Along similar lines is a story told to me by Wolf Biermann which is unworthy even of Colonel Ehrhardt of the Gestapo in *To Be or Not to Be*. In the early 1990s he met with green political groups in East Germany; among them there were some neo-Nazi ecologists, and when Biermann reproached them for their sympathy for Hitler, he got a shocking reply: 'No, we are deeply critical of Hitler. True, he did some good things, like getting rid of the Jews, but he also did many horrible things, like destroying forests to build the highways.'

(Note how this critique inverts the usual defence of Hitler: 'True, he did some horrible things like killing the Jews, but he also did some good things like building highways and making trains run on time!')

Lubitsch's approach has a deep ontological foundation. In one of the most effective jokes in *To Be or Not to Be*, the Polish actor Josef Tura impersonates Colonel Ehrhardt in conversation with a high-level Polish collaborator and, in (what we take as) a ridiculously exaggerated way, he comments on rumours about himself, 'So they call me Concentration-Camp-Erhardt?' and accompanies his words with vulgar laughter. A little bit later, Tura has to escape and the real Ehrhardt arrives; when the conversation again touches on rumours about him, he reacts in exactly the same ridiculously exaggerated way as his impersonator. The message is clear: even Ehrhardt himself is not immediately himself, he also imitates his own copy or, more precisely, the ridiculous idea of himself. While Tura acts him, Ehrhardt acts himself. (Incidentally, here we have a perfect example of the Hegelian distinction between subjective and objective humour: Tura playing Ehrhardt in an exaggerated way is subjective humour, with Tura making fun of him, while Ehrhardt enacting the same exaggeration is objective humour, humour inscribed into the object itself.) Could we not say exactly the same of Donald Trump, who also acts himself?

This is not to say that Lubitsch is a postmodern cynical ironist whose premise is that, since everything is mediated and indirect, with each of us playing ourselves, true love exists in some romantic sphere beyond the comic indirectness. We have to learn to locate it amidst all these comic confusions. If there is an example of true and permanent love in *To Be or Not to Be*, a model of an ideal marriage, it is that of Josef and Maria Tura (Joseph and Mary, the ultimate couple!). Maria is constantly flirting around and cheating on him, while Josef is intolerably self-centred and convinced of his greatness, but they are totally inseparable. One cannot even imagine their divorce – there is no question of Maria dropping Josef and going to live with the pilot with whom she cheats on him. What this means is that there is no universal formula for a successful sexual relationship: the only universality is a negative one, that of failure, and, to offset this failure, a couple should invent an idiosyncratic formula – what Lacan called *sinthom*, the minimal knot of enjoyment – which, if it works, can be much more stable than pure, passionate love.

But, again, does this fact not also point to the limitations of Lubitsch's approach for us today? Increasingly we are experiencing how what for Lubitsch was still a joke is now simply enacted in real (political and ideological) life. Recall Ehrhardt's legendary quip, 'We do the concentrating, Poles do the camping' – could not today's managers advocating the politics of austerity be saying something similar? 'We do the politics, ordinary people do the austerity.' Maybe, Lubitsch-type jokes only work while we still have liberal hypocrisy to mock; but what about when power exerts itself brutally, discarding its liberal-humanitarian-democratic mask? One is almost tempted to say: bring back the hypocritical mask!

However, Lubitsch would have been aware that such sudden removal of the mask is always fake. In the 'revolutionary' 1960s, it was fashionable to assert perversion against the compromise of hysteria: a pervert directly violates social norms, he or she does openly what a hysteric only dreams about or articulates ambiguously in his or her symptoms. That is to say, the pervert effectively moves beyond the Master and his Law, while the hysteric merely provokes her Master in an ambiguous way which can also be read as a demand for a more authentic Master ... Against this view, Freud and Lacan consistently emphasized that perversion, far from being subversive, is the hidden obverse of power: every form of power needs perversion as its inherent transgression which sustains it. In order to be operative, every ideological edifice has to be inconsistent: its explicit norms have to be supplemented by higher-level implicit norms which tell us how to deal with those explicit norms (when to obey them and when to violate them). In other words, an ideology does not just consist of its explicit norms; it always comprises an obscene underside which violates those explicit norms – this inconsistency is what makes it an ideology. However, what is happening today is not just more of the same, but a qualitatively new form of dissonance: a dissonance openly admitted, and for that reason treated as irrelevant. The story about the ashtray with which we began this book provides the matrix for this new form – recall the debates on torture: was the stance of the US authorities not something like, 'Torture is prohibited, and here is how you do waterboarding'? The paradox is thus that, today, there is in some ways less deception than in a more traditional functioning of ideology: nobody is really deceived.

It is precisely when we appear to open ourselves up to the dirtiest fantasies of our mind that the truly traumatic point remains repressed. However, isn't Lubitsch's indirectness also conditioned by the Hays Code censorship? Adorno wrote somewhere that a really good film would follow all the rules of Hays Code, although not in order to obey the law but out of an immanent necessity – this is what Lubitsch is doing, but not quite . . .

An exemplary case of how these rules work is found in the well-known brief scene three quarters into *Casablanca*: Ilsa Lund (Ingrid Bergman) comes to Rick Blaine's (Humphrey Bogart's) room to try to obtain the letters of transit that will allow her and her Resistance leader husband Victor Laszlo to escape from Casablanca to Portugal and then to America. After she breaks down and says, 'If you knew how much I loved you, how much I still love you,' they embrace in close-up which then dissolves into a three-and-a-half-second shot of the airport tower at night, its searchlight circling, and then back to a shot from outside the window of Rick's room, where he is standing, looking out and smoking a cigarette. He turns into the room and says, 'And then?' She resumes her story . . . The question that immediately pops up here is, of course: what happened *in between*, during that shot of the airport – did they have sex or not? The film is not simply ambiguous; rather it generates two very clear, although mutually exclusive, meanings – they did it, and they didn't do it. It gives a series of codified signals that they did do it, and that the three-and-a-half-second shot stands for a longer period of time (the sight of a couple passionately embracing usually signals the sexual act after the fade-out; the cigarette afterwards is also the standard signal of relaxation after the act; and there is the vulgar phallic connotation of the tower). A parallel series of signals suggests that they did not do it, that the shot of the airport tower corresponds to the real diegetic time (the bed in the background is undisturbed; the same conversation seems to go on without a break). While, at the level of its surface storyline, the film can be constructed by the spectator as obeying the strictest moral codes, it simultaneously offers to the sophisticated sufficient clues to construct an alternative, sexually much more daring narrative. This is how ideology works in classic Hollywood: nothing is totally repressed, everything can be unambiguously signalled in a codified way (if someone remarks that a guy smells of perfume, it means he is gay, etc.).

In later Hollywood, this game of inherent transgression got much more complex. Recall what is arguably the most powerful scene in *The Sound of Music* (1965): after Maria escapes from the von Trapp family back to the monastery, unable to deal with her sexual attraction towards Baron von Trapp, she cannot find peace there since she is still longing for him; in a memorable scene, the Mother Superior summons her and advises her to return to the von Trapp family and try to sort out her relationship with the Baron. She delivers this message in a weird song, 'Climb every mountain!', whose surprising motif is: do it! Take the risk and try everything your heart wants! Do not allow petty considerations to stand in your way! So the very person who one would expect to preach abstinence and renunciation turns out to be the agent of fidelity to one's desire. Significantly, when *The Sound of Music* was shown in (still socialist) Yugoslavia in the late 1960s, *this* scene – the three minutes of this song – was the only part of the film which was censored. The anonymous socialist censor thereby displayed his profound sense of the truly dangerous power of Catholic ideology: far from being the religion of sacrifice, of renunciation of earthly pleasures (in contrast to the pagan affirmation of the life of passions), Catholicism offers a devious stratagem to indulge in our desires *without having to pay the price for them*, to enjoy life without fear of the decay and debilitating pain awaiting us at the end of the day. Today, with cases of paedophilia popping up all around in the Catholic Church, one can easily imagine a new version of the scene: a young priest approaches the abbot, complaining that he is still tortured by desires for young boys, and demanding further punishment; the abbot answers by singing, 'Climb every young boy . . . '

Lubitsch, however, is not doing this: his indirectness does not amount to such a primitive game, where precise codes signal what happens behind closed doors (the sexual act or something similar). Lubitsch is well aware that such a technique perversely supplements the law with its obscene underside: the perverse direct enactment of the repressed content equals the strongest repression, i.e. is precisely when we appear to open ourselves up to the dirtiest fantasies of our mind that the truly traumatic point remains repressed. So what is Lubitsch doing?

Although I am not a fan of *Sex and the City*, an interesting point is made in one of the episodes where Miranda gets involved with a guy who likes to talk dirty during sex, and since she prefers to keep silent, he asks her also to voice whatever dirty things pop up in her mind, with no restraint. At first she resists, but then she too gets caught up in this game, and things work well, their sex is intense and passionate, till . . . till she says something that really disturbs her lover and makes him totally withdraw into himself and leads to the breakdown of their relationship. In the middle of her babble she mentions that she has noticed how he enjoys it when, while he makes love to her, he pushes her finger into his ass. Unknowingly, she thereby touches on the exception: yes, talk about anything you want, spill out all the dirty images that come into your head, *except that*. The lesson of this incident is important: even the universality of talking freely is based on some exceptions other than extreme brutality. The prohibited detail is in itself a minor and rather innocent thing, and we can only guess why the guy is so sensitive about it – in all probability, it is because the passive experience it involved (anal penetration) disturbs his masculine identification. The detail disturbed him not because it was simply too much for him, but because it touched upon his innermost phantasmatic kernel which he was unable to confront openly, a *sinthom* of (a knot that condenses) his enjoyment. We can imagine Miranda asking him what he wants her to do to him during their love-making, and we can be sure he would never be able to mention *that*, the thing he most desires – it is left to her to discover that while keeping silent about it. One should also note how the misunderstanding between Miranda and her lover follows Lacan's formula of sexuation: the lover understands his request in 'masculine' terms of universality based on an exception (talk dirty and say everything . . . except *that*), while Miranda understands it in 'feminine' terms of non-all with no exception (just tell whatever comes to your mind without worrying if you say it 'all' and therefore without any exceptions). Elena Ferrante recently wrote: 'Even today, after a century of feminism, we can't fully be ourselves.' But what if the very idea of 'being fully oneself' is a masculine idea?[1]

The German philosopher F. W. J. Schelling defined 'uncanny' as the appearance, the coming out into the open space, of something that should remain hidden. This is how the uncanny causes anxiety – not because it confronts us with the fact that something is lacking, but because lack itself is lacking, because we get too much. Miranda's lover feels castrated because he gets too much from her, more than he really asked for – he asked her to voice all the obscenities that pop up in her mind, and what he gets is the exception on which his universality was based. He experiences castration here – not lack, but this 'too much' is castrating.

In Lubitsch's *The Smiling Lieutenant* of 1931 (just prior to the imposition of the Hays Code), this obscene excess is brought to the extreme.[2] The first five minutes of the film enact the passage from indirectness (the basic operation of the 'Lubitsch touch') to excess (its inherent obverse). It begins with a short scene of indirectness: a formally dressed person climbs up the stairs, stops in front of an apartment door, pulls out of his briefcase a document (a bill for expensive clothes) and rings the bell; nobody answers, so he knocks on the door, but, again, nobody answers, and he leaves. Immediately after his departure, a young lady comes up the stairs and also knocks on the door, but this time her knocks follow an obviously coded pattern. The door opens and she enters; after an interlude signalled by the turning-off of lights, she exits the door, full of joy. In this indirect way, we learn all the essentials about the inhabitant of the apartment without even seeing him: he is Niki, an officer in the Imperial Austrian Army (as the plate near the door informs us), who likes to dress expensively and enjoy life, and who serially seduces girls. But now comes the counter-scene: immediately after the girl leaves, we cut to the interior of the apartment and see Niki (played by Maurice Chevalier) in a nightgown; also satisfied, he stands up, confronts us, the spectators, and sings an extremely obscene and embarrassing song praising army life. The song is based on the parallel between military exercises (following orders, attacking, shooting) and love-making. It describes the officer's duty as that of 'shooting down' girls, and Niki enacts it with the suggestively emphasized word 'ratatatata'; what adds to the obscenity is that Chevalier's performance is also done in his French accent, full of French words, drawing on the common popular-culture image of the sophisticatedly seductive and promiscuous Frenchman (and totally inconsistent with the fact that he presents himself as an Austrian officer). The parallel between military and sexual activity also emphasizes the point that, in copulating, a man is serving/servicing a woman, obeying her orders, which is why when, after his marriage to the princess, Niki deliberately pretends not to understand her sexual invitations and refuses to oblige, we should read it as an act of rebellion, a worker refusing to serve his master.

A series of comical reversals follow, and, at the film's end, after Niki allows himself to be seduced by the princess, he again exits the bedroom door and addresses us with the same song, just with slightly changed words (praising not mere passing affairs but marital sex), ending once more with the obscene 'ratatatata'. When he opens the bedroom door, we hear from the inside the voice of his wife repeating 'ratatatata'. 'The thing' (the sexual act) still happens behind the door, so that, at a formal level, indirectness prevails; but the obscenity of what goes on in front of the door (the song with its 'ratatatata' of shooting at – ejaculating into – a woman) is in some sense much more 'dirty' than a direct depiction of what went on behind that closed door. Back to Miranda from *Sex and the City*: 'ratatatata' plays exactly the same role as 'sticking the finger into the asshole', that of a detail which should have remained unspoken, hidden. The direct obscenity of this 'ratatatata' episode indicates that *The Smiling Lieutenant* was shot in the pre-Hays-Code era, and also before Lubitsch established his 'touch': in mature Lubitsch, such direct obscenity is excluded. In *The Smiling Lieutenant* (1931) (and, as we shall see, in *Broken Lullaby*, 1932), we have Lubitsch elements, but they are isolated, in their raw state – 'out of touch'.

Against Contractual Sex

This is why Lubitsch would have been horrified at the idea of the sexual contracts that are proliferating in the aftermath of the MeToo movement, from the US and the UK to Sweden. Their declared goal is, of course, clear: to exclude elements of violence and domination in sexual contact. The idea is that, before having sex, both partners should sign a document stating their identity, their consent to engage in sexual intercourse, as well as the conditions and limitations of their activity (use of condom, of dirty language, the inviolable right of each partner to step back and interrupt the act at any moment, to inform his/her partner about health issues, religion, and so on). It sounds good, but a series of problems and ambiguities arise immediately.

Let's begin with the basics. In the West, at least, we are becoming massively aware of the extent of coercion and exploitation in sexual relations. However, we should bear in mind also the no less massive fact that on a daily basis millions of people flirt and play the game of seduction, with the clear aim of attracting a partner with whom to make love. The result in modern Western culture is that both sexes are expected to play an active role in this game. When women dress provocatively to attract male looks, when they 'objectify' themselves to seduce them, they don't do it by offering themselves as passive objects: they are the active agents of their own 'objectification', manipulating men, playing ambiguous games, including the full right to step out of the game at any moment even if, to the man, it appears to contradict previous 'signals'. This active role of women is their freedom, which so bothers all kinds of fundamentalists, from Muslims who recently prohibited women from touching and playing with bananas and other fruit resembling the penis to our own ordinary male chauvinist, who explodes in violence against a woman who first 'provokes' him and then rejects his advances. Female sexual liberation is not just a puritanical withdrawal from being 'objectivized' (as a sexual object for men), but the right actively to play with self-objectivization, offering oneself and withdrawing as one wishes. Will it be still possible to state these simple facts, or will politically correct pressure compel us to back up these games with some formal-legal proclamation of consensuality?

Yes, sex is traversed by power games, by violent obscenities, but the difficult thing to admit is that they are inherent in it. The problem is that sexuality, power and violence are much more intimately intertwined than we expect, so that elements of what is considered brutality can also be sexualized, that is, libidinally invested – after all, sadism and masochism are forms of sexual activity. Sexuality purified of violence and power games can well end up being desexualized. Some perceptive observers have already noticed how the only form of sexual relation that fully meets politically correct criteria would be a contract drawn up between sado-masochistic partners. The rise of political correctness and the rise of violence are thus two sides of the same coin: insofar as the basic premise of political correctness is the reduction of sexuality to contractual mutual consent, Jean-Claude Milner was right to point out how the gay rights movement unavoidably reaches its climax in contracts which stipulate extreme forms of sado-masochistic sex (treating a person like a dog on a collar, slave-trading, torture, up to consented killing). In such forms of consensual slavery, the market freedom of a contract negates itself: slave-trading becomes the ultimate assertion of freedom. It is as if the motif of 'Kant with Sade' becomes a reality in an unexpected way.

So, how to combat this tendency? The first task is to make sure that the ongoing explosion of feminist struggle will not remain limited to the public life of the rich and famous but will trickle down and penetrate the daily lives of millions of ordinary 'invisible' individuals. And the last (but not least) point is to explore how to link this awakening to ongoing political and economic struggles, how to prevent it from being appropriated by Western liberal ideology and practice as yet another way to reassert its supremacy. Remember how many of the accused, beginning with Harvey Weinstein, reacted by publicly proclaiming that they will seek help in therapy – a disgusting gesture if ever there was one! Their acts were not cases of private pathology, they were expressions of the predominant masculine ideology and power structures, and it is the latter that should be changed.

At approximately the same time as the Harvey Weinstein scandals began to roll, the Paradise Papers were published, and one cannot but wonder why nobody demanded that people should stop listening to the songs of Bono (the great humanitarian, always ready to help the poor in Africa) or of Shakira because of the way they avoided paying taxes and thus cheated the public authorities of large sums of money, or that the British royal family should receive less public money because they parked part of their wealth in tax havens, while the fact that Louis CK showed his penis to some ladies instantly ruined his career. Isn't this a new version of Brecht's old motto, 'What is robbing a bank compared to founding a bank?'? Cheating with big money is tolerable while showing your penis to a couple of people instantly makes you an outcast?[3] This is why contracts will never really work. Should sexual contracts be legally binding or not? If not, what prevents brutal men just signing one and then violating it?

If yes, can one even imagine the legal nightmare its violation may involve? This does not mean that we should endorse the letter signed by Catherine Deneuve and others which criticizes the 'excesses' of MeToo 'puritanism' and defends traditional forms of gallantry and seduction. The problem is not that MeToo goes too far, sometimes approaching a witch hunt, and that more moderation and understanding are needed, but the way MeToo addresses the issue. In downplaying the complexity of sexual interaction, it not only blurs the line between lewd misconduct and criminal violence but also cloaks invisible forms of extreme psychological violence as politeness and respect.

The Rotherham scandal (Pakistani youth gangs terrorizing and serially raping hundreds of white girls from poor areas) is now repeated in Telford and some other British cities. The Left again demonstrates its inability to openly confront this problem in a non-racist way – it prefers to relativize or minimalize it in order not to fall into Islamophobia. They are obviously not aware that every time we find excuses to avoid this topic, we bring new votes to the alt-right. And where are the MeToo feminists here? Sometimes it looks as if they care more about a couple of affluent women who were shocked when Louis CK showed them his penis than when hundreds of poor girls are being brutally raped.

Furthermore, the scene described again and again by the MeToo partisans is that of a predatory man threatening to rape a woman (or at least coerce her into sex) – but what about the majority of women who (like men) desire to have sex but are ignored for not being attractive enough? Can one imagine their suffering, especially in our PC times, when 'a beautiful woman' is more and more considered a male-predatory phrase objectivizing women (and, of course, the perception that some women are beautiful and attractive persists even stronger at the unspoken level . . .)?

In replying to those who insisted on a difference between Weinstein and Louis CK, MeToo activists claimed that those who say this have no idea how male violence works and is experienced, and that masturbation in front of women can have no less violent an effect than male physical force. Although there is an element of truth in these claims, nonetheless a clear limit should be imposed to the logic that sustains this argumentation: what one feels cannot be the ultimate measure of authenticity, since feelings can also lie – if we deny that, we simply deny the Freudian unconscious. (Incidentally, this reference to feeling as the ultimate criterion of authenticity faithfully reproduces the old anti-feminist prejudice, elaborated, among others, by Descartes and other early rationalists, about women as beings who are totally determined by their emotions and cannot rise above them through reflection.) In truly effective patriarchal domination, a woman doesn't even experience her role as being that of a humiliated and exploited victim; she simply accepts her submission as part of the order of things.

There is an obvious pop-cultural reply to the thesis on the ongoing male oppression of and domination over women – E. L. James's mega-bestseller *Fifty Shades of Grey*, a novel written by a woman about a woman who enjoys her sexual submission to a men, and (so the media tell us) wildly popular among women. In answering this critical point, one should, of course, avoid at any price fast pseudo-psychoanalytic counter-claims in the style of 'James's novel makes it clear that even women who appear to demand emancipation from male power are effectively in the thrall of a profound unconscious masochistic desire to be dominated by men.' The feminist claim that women engaging in masochist fantasies present a case of identifying with the enemy and internalizing the patriarchal viewpoint fares no better. The first thing to do is to take a closer look at what *Fifty Shades of Grey* does: it does not imply enjoying actual subordination but rather enjoying a *fantasy* of subordination, which is absolutely not the same and should in no way be interpreted as a call for actual subordination. One of the basic lessons of psychoanalysis is that, when our innermost fantasies are imposed on us from outside, the experience is utterly devastating – to put it bluntly, when a woman who secretly dreams about being treated roughly during sex is effectively raped, the effect is much more brutal than in the case of a rape not echoing such fantasies.

A further point to be made (and which was amply developed by Deleuze) is that there is no symmetry between masochism and sadism: while a sadist brutally mistreats his victim in order to humiliate him/her, masochism relies on a contract which stipulates the exact terms of the interplay, inclusive of the limits of violence (which is as a rule theatrically staged). Is this not what also happened in *Fifty Shades*? The two partners conclude a contract out of which they are free to step at any moment. Also, the violence enacted is very gentle – no comparison is possible here with the actual misery of women terrorized by their partners. (There is, of course, a different kind of actual feminine masochism, but this is emphatically not what *Fifty Shades* is dealing with.) In some sense, one could even claim that such a masochist contract presents a case of feminine empowerment: it is the woman who installs a man in the theatrical role of her Master and defines the terms of their interaction. This is what Lacan meant by his answer to Freud's question, 'What does a woman want?' – a master, but a master whom she can dominate and manipulate.

The standard form of the masochist contract – recall Sacher-Masoch's *Venus in Furs* – has the male partner in the position of the 'victim' who contractually installs a woman as his Domina and instructs her in precise terms what to do to him (whip him, step on him, humiliate him with vulgar words, etc.). If we are to believe reports in the media, such contracts are popular among top managers, who supplement their brutal exercise of administrative authority with enacting masochistic fantasies – this in no way diminishes their actual social power, it just functions as its obscene supplement. Is the fact that, in *Fifty Shades*, a woman (not just the novel's heroine but also the writer and the vast feminine public) takes over this role not a twisted sign of the decline of patriarchy? One of the definitions of a Master is precisely 'the one who has the right to enact his/her fantasies'.

But violence enters sexuality at an even more elementary level. When somebody unexpectedly declares passionate love to us – is not the first reaction, preceding the possible positive reply, that something obscene, intrusive, is being forced upon us? In the middle of Alejandro Inarritu's *21 Grams*, Paul, who is dying of a weakened heart, softly declares his love to Cristina, who is traumatized by the recent death of her husband and two young children, and then quickly withdraws; when they next meet, Cristina explodes into a complaint about the violent nature of declaring love:

> You know, you kept me thinking all day. I haven't spoken to anyone for months and I barely know you and I already need to talk to you . . . And there's something the more I think about the less I understand: why the hell did you tell me you liked me? . . . Answer me, because I didn't like you saying that at all . . . You can't just walk up to a woman you barely know and tell her you like her. Y-o-u-c-a-n't. You don't know what she's going through, what she's feeling . . . I'm not married, you know. I'm not anything in this world. I'm just not anything.[4]

Upon this, Cristina looks at Paul, raises her hands and desperately starts kissing him on the mouth; so it is not that she did not like him and did not desire carnal contact with him – the problem for her was, on the contrary, that she *did* want it, i.e., the point of her complaint was: what right does he have to stir up her desire? So, again, even when sexual contact is desired by both parts, there can be an element of violence in initiating it, the violence of, precisely, initiating it in a direct way. The reason is simply that sexual desire never fits the image of one's self – it is always experienced as a violent intrusion. No contracts help here – demanding a contract can be in itself a form of violence (which, in special circumstances, can again become part of a masochist sexual game). And this is what complicates any direct attempt to regulate things. When commentators try to resume the results of the ongoing new wave of the struggle for women's emancipation, one of their conclusions is that 'no means no' is not enough to lead a 'happy sex life' since it still leaves the space open for more subtle forks of coercion – here is an exemplary case of this line of argumentation:

> Badgering someone into queasy submission might technically be within the law, but it is not the road to a happy sex life and it may no

> longer protect a man from public censure. What young men should look for, Tillman argues, is not the potentially ambiguous absence of 'no', but the enthusiastic presence of a 'yes, yes, yes' or affirmative consent. 'In 2018, "no means no" is totally antiquated. It puts all the pressure on the person in the most vulnerable position, that if someone doesn't have the capacity or the confidence to speak up, then they're going to be violated,' she says. 'If somebody isn't an enthusiastic yes, if they're hesitating, if they're like, "Uh, I don't know" – at this point in time, that equals no.'[5]

One cannot but agree with all the critical points in this passage: how a weak 'yes' under pressure equals 'no', etc. What is problematic is 'the enthusiastic presence of a "yes, yes, yes" ' – it is easy to imagine in what a humiliating position this condition can put a woman who, to put it bluntly (and why not?), passionately wants to get laid by a man – basically, she has to perform an equivalent of publicly stating 'Please fuck me!' . . . Are there not much more subtle (but nonetheless unambiguously clear) ways to do this? Furthermore, if one looks for 'the road to a happy sex life', one looks for it in vain for the simple reason that there is no such thing: things always, for immanent reasons, go wrong in some way in sex, and the only chance of a relatively 'happy sex life' is to find a way to make these failures work against themselves. Directly searching for 'the road to a happy sex life' is the safest way to ruin things, and the imagined scene of both partners enthusiastically shouting 'yes, yes, yes' is in real life as close as one can get to hell.

Things get even more complex with the right to withdraw from sexual interaction at any moment – one rarely mentions how this right opens up new modes of violence. What if the woman, after seeing her partner naked with an erect penis, begins to mock him and tells him to leave? What if the man does the same to her? Can one imagine a more humiliating situation? The extreme case of the violence of such withdrawal is the painful scene from David Lynch's *Wild at Heart* in which Bobby Peru (played by Willem Dafoe) violently imposes himself on Lula (Laura Dern), intruding into her space and obscenely whispering and shouting at her: 'Say fuck me!' When, after painfully protruded pressure, she concedes and whispers 'Fuck me' (in an ambiguous way where coercion is inextricably mixed with inner arousal), he steps back and tells her with a smile: 'Not now, I gotta go! But another time, gladly.' The effect is so humiliating to her that, in some sense, the symbolic violence of this withdrawal, of the rejection of the enforced offer, is worse than if he were to accept her offer and actually fuck her. Clearly, one can find an appropriate way to resolve such impasses only through manners and sensitivity, which by definition cannot be legislated for. If one wants to prevent violence and brutality by adding new clauses to the contract, one loses a central feature of sexual interplay, which is precisely a delicate balance between what is said and what is not said. Sexual interplay is full of such exceptions, where a silent understanding and tact offer the only way to proceed when one wants things done but not explicitly spoken about, when extreme emotional brutality can be enacted in the guise of politeness and when moderate violence itself can get sexualized. If we go to the end on this path, we have to conclude that even an enthusiastic 'yes, yes, yes' can effectively function as a mask of violence and domination. Monica Lewinsky recently said that

> she stands by her 2014 comments that their relationship was consensual, but muses about the 'vast power differentials' that existed between the two. Ms Lewinsky says she had 'limited understanding of the consequences' at the time, and regrets the affair daily. 'The dictionary definition of "consent"? To give permission for something to happen,' she wrote. 'And yet what did the "something" mean in this instance, given the power dynamics, his position, and my age? . . . He was my boss. He was the most powerful man on the planet. He was 27 years my senior, with enough life experience to know better.'[6]

True, but she did not just consent, she directly initiated sexual contact, and it was Clinton who 'consented', and the 'vast power differential' was probably a key part of his attraction for her. As for her claim that, since he was an older, more experienced man, he should have 'known better' and have rejected her advances, is there not something hypocritical in this self-ascribed role of an inexperienced victim? Do we not find ourselves here at the exact, almost symmetrical, opposite of the Muslim fundamentalist view according to which a man who rapes a woman has been secretly seduced (provoked) by her into doing it? Such a reading of male rape as the result of woman's provocation is often reported by the media. In the autumn of 2006, Sheik Taj Din al-Hilali, Australia's most senior Muslim cleric, caused a scandal when, after a group of Muslim men had been jailed for gang rape, he said, 'If you take uncovered meat and place it outside on the street . . . and the cats come and eat it . . . whose fault is it – the cat's or the uncovered meat? The uncovered meat is the problem.' The explosively scandalous nature of this comparison between a woman who is not veiled and raw, uncovered meat distracted our attention from another, much more surprising premise underlying al-Hilali's argument: if women are held responsible for the sexual conduct of men, does this not imply that men are totally helpless when faced with what they perceive as a sexual provocation, that they are simply unable to resist it, that they are totally enslaved to their sexual hunger, precisely like a cat when it sees raw meat? In contrast to this presumption of the complete lack of male responsibility for their own sexual conduct, the emphasis on public female eroticism in the West relies on the premise that men *are* capable of sexual restraint, that they are not blind slaves of their sexual drives.

This total responsibility of the woman for the sexual act strangely mirrors the Lewinsky view that, although the initiative was fully on her side, the responsibility was fully Clinton's. In the same way that, in the Muslim fundamentalist view, men are helpless victims of woman's perfidious seduction even if they commit a brutal rape, in the Lewinsky case, she was a victim even if she provocatively initiated the affair. The symmetry between the two cases is flawed, of course, since in both of them men are in the actual position of social power and domination. However, playing the card of a helpless victim in such a case as Lewinsky's is a self-humiliating spectacle which in no way helps women's emancipation – it merely confirms man as the master.

Those who admit the existence of so-called 'grey zones' (between the two extremes of mutually desired sexual interplay and clear violent imposition) as a rule miss their changing status within one and the same sexual interplay. Especially today, in our politically correct times, a seduction process always involves the risky move of 'making a pass' – at this potentially dangerous moment, one exposes oneself, one intrudes into another person's intimate space. The danger resides in the fact that, if my pass is rejected, it will appear as a Politically Incorrect act of harassment; so there is an obstacle I have to overcome. Here, however, a subtle asymmetry enters: if my pass is accepted, it is not that I have successfully overcome the obstacle – what happens is that, retroactively, I learn that *there never was an obstacle to be overcome*.

One should also bear in mind that patriarchal domination corrupts both of its poles – or, to quote Arthur Koestler: 'If power corrupts, the reverse is also true: persecution corrupts the victims, though perhaps in subtler and more tragic ways.' Consequently, one should also talk about female manipulation and emotional brutality (ultimately as a desperate reply to male domination): women fight back any way they can. And one should admit that, in many parts of our society in which traditional patriarchy is to a large extent undermined, men are no less under pressure, so the proper strategy should be to address male anxieties too and to strive for a pact between women's struggle for emancipation and male concerns. Male violence against women is largely a panicky reaction to the fact that traditional male authority is enfeebled, and part of the struggle for emancipation should be to demonstrate to men how the acceptance of emancipated women will release them from their anxieties and enable them to lead more satisfied lives.

In a recent polemical exchange, some feminists reacted to Germaine Greer's critical remarks on MeToo;[7] their main point was that, while Greer's thesis – that women should sexually liberate themselves from male domination and assume an active sexual life without any recourse to victimhood – was valid in the sexual-liberation movement of the 1960s, today the situation is different. What happened in between is that the sexual emancipation of women (their involvement in social life as active sexual beings, with full freedom of initiative) was itself commodified: true, women are no longer perceived as passive objects of male desire, but their active sexuality itself now appears in male eyes as their permanent availability, their readiness to engage in sexual interaction. In these new circumstances, forcefully saying 'NO!' is not mere self-victimization, since it implies the rejection of this new form of sexual subjectivization of women, of the demand that women not only passively submit to male sexual domination but behave as if they actively want it.

While there is a strong element of truth in this line of argumentation, we should nonetheless add at least two points. We must never forget that this false subjectivization is sustained by internalized superego pressure, so the first step for women is to liberate themselves from this pressure – as Herbert Marcuse put it back in the 1960s, freedom is a condition of liberation: in order to liberate yourself, you first have to break the inner chains of ideology. Second, there is still a big divide between displaying active sexuality in order to please male desire and effectively acting as autonomous sexual agents – the latter does not please prospective male partners and tends to trigger anxiety in them.

This brings us back to contractual sex: what makes it problematic is not only its legal form but also its hidden bias. It obviously privileges casual sex, where partners don't yet know each other and want to avoid misunderstandings about their one-night stand. We need to extend our attention also to the long-term relationship permeated with forms of violence and domination in much more subtle ways than spectacular Weinstein-style enforced sex. To paraphrase Brecht yet again, what is the fate of a movie star who was once blackmailed into sex (or directly raped) to ensure her career compared with a miserable housewife who is for long years constantly terrorized and humiliated by her husband?

Ultimately, no laws and contracts help here, only a revolution in mores; and this brings us back to Lubitsch's key procedure, his famous indirectness, his refusal blatantly to portray the thing itself, sex or violence. It is a powerful technique: ultimately, it means that the couple is never alone in sex, that a third element is always implied, even if it is just the imagined gaze of a witness. The clearest case is his *One Hour with You* (1932), in which a woman and a man, Mitzi and Andre, each married to her or his partner, accidentally find themselves in the same taxi. Their affair is set in motion by the fact that, to an imagined observer, it seems as if they are lovers, although they are just sitting in the privacy of a car. Mitzi says: 'Look at us: he reads a newspaper, she looks through the window . . . hahaha . . . ' Then she adds in a more serious vein: 'Try to explain this to your wife!' Andre cannot resist the power of appearance; although he clearly loves his wife, the way this scene looks is incriminating and its effects cannot be erased. Mitzi does not refer here primarily to an instance which is physically present, in the sense of 'if someone actually sees us now, he or she will automatically conclude that it is a love affair'. The instance she has in mind is much more complex and calls for the notion deployed among others by Robert Pfaller: that of a naive observer who does not judge the situation with regard to the subject's true intentions but exclusively according to how things appear.[8]

The topic of a decentred fantasy which sustains a sexual relationship takes a weird turn in Lubitsch's *Broken Lullaby* (1932), which is often dismissed as a failure, but it brings out this feature of his work isolated from the context of mature Lubitsch – out of touch, as it were, not yet as part of Lubitsch touch. Here is the outline of the story. Haunted by the memory of Walter Holderlin (!), a soldier he killed during the First World War, the French musician Paul Renard travels to Germany to find Walter's family, using the address on a letter he found on the dead man's body. Dr Holderlin, Walter's father, initially refuses to welcome Paul into his home, but changes his mind when his son's fiancée, Elsa, identifies him as the man who has been leaving flowers on Walter's grave. Rather than reveal the real connection between them, Paul tells the Holderlin family that he was a friend of their son, who attended the same musical conservatory. Although the hostile townspeople and local gossips disapprove, the Holderlins befriend Paul, who falls in love with Elsa; after some prevarication, he tells her the truth about killing Walter. She convinces him not to confess to Walter's parents, who have embraced him as their second son, and Paul agrees to forgo easing his conscience and stays with his adopted family. Dr Holderlin presents Walter's violin to Paul and, in the film's final scene, Paul plays the violin while Elsa accompanies him on the piano, observed in the loving gazes of the parental couple. There is something disturbing about the film, a weird oscillation between poetic melodrama and obscene humour. The couple (the young woman and the killer of her previous fiancé) are happily united, under the protective gaze of her dead fiancé's parents – it is this gaze that provides the fantasy frame for their relationship, and the obvious question is: do they really act as lovers just for the sake of the parents, or is this gaze an excuse for them to engage in sex? This question is, of course, a false one, because it doesn't matter which of the alternatives is true: even if the parents' gaze is just an excuse for sex, it is still a necessary excuse.

Sometimes, real life catches up with Lubitsch, taking his plot in a way which pushes things a little bit further. The basic situation of his *The Shop Around the Corner* (1940) occurred in real life in (of all places) Sarajevo in the mid-1990s, just after the siege of the city. A young married couple were in crisis, the husband and wife getting bored of each other, so in order to revitalize their emotional life, each of them gets engaged in internet flirting with an anonymous partner, exchanging dreams with him or her, and so on. Since, in both cases, it appeared that they had each found their ideal partner, they both decided to meet in reality. When they meet in a cafeteria, they are shocked to discover that they have dated each other, husband and wife. So what is the lesson of this coincidence? Did it lead them to discover the inner harmony of their dreams and thus make them stay together with a deeper understanding? I think Lubitsch would have been more inclined to see such a proximity of their inner dreams as a bad omen, and would predict that they would run away from each other in horror . . .

And, again, this indirectness is at work on every level – I think Lubitsch would not have been surprised to learn that the paradigmatic hardcore sexual position (and shot) is that of the woman lying on her back with her legs spread wide backwards and her knees above her shoulders; the camera is in front, showing the man's penis penetrating her vagina (the man's face is as a rule invisible; he is reduced to an instrument), but what we see in the background between her thighs is her face in the thrall of orgasmic bliss. This minimal 'reflexivity' is crucial: if we were just to see the close-up of penetration the scene would soon become boring, disgusting even, more of a medical showcase – one has to add the woman's enthralled gaze, the subjective reaction to what is going on. Furthermore, this gaze is not directed at her partner, but at us, the viewers, confirming to us her pleasure – we, the spectators, clearly play the role of the big Other who has to register her enjoyment. The pivot of the scene is thus not male (her sexual partner's or the spectator's) satisfaction – the spectator is reduced to a pure gaze: it is the woman's sensual gratification (staged for the male gaze, of course). The irony here is that the very fact that the woman is not 'objectivized' but rendered as a subject makes her humiliation worse. This elementary hardcore scene perfectly renders the minimal reflexivity that cuts from within every immediate orgasmic One.

This is why we don't just 'do it', but we have to make love. In a nice dialogue sequence in Alexander Payne's *Downsizing* (2017), after the hero has sex with a Vietnamese refugee lady, she asks him in her imperfect English whether this was for him a love fuck, just a sex fuck, a mercy fuck, and so on, and in reply he asks why she uses the vulgar 'F' word instead of the more polite and gentle 'make love'. She takes the point and goes on to talk about 'make fuck' instead of just 'fuck' - 'Why did you make fuck to me?' And in a way, she is right: maybe the definition of love is that you don't just fuck your partner, you make fuck to him or her . . .

Cynicism, Humour and Engagement

Here, however, we encounter Lubitsch's basic ambiguity. Is his solution, his third way, not the benevolent hedonism of indirectness, whereby we enjoy all the detours which eroticize sex? To clarify this important point, let us confront the key question: where is the trouble in paradise in *Trouble in Paradise* (1932)? The lyrics of a song heard during the credits provide a clue to the 'trouble' alluded to (as does the image that accompanies the song: first we see the words 'trouble in', then beneath them a large double bed appears, and then, over the surface of the bed, in large letters, 'paradise'). So 'paradise' is that of a full sexual relationship: 'That's paradise / while arms entwine and lips are kissing, / But if there's something missing / that signifies / trouble in paradise.' To put it in a brutally direct way, 'trouble in paradise' is thus Lubitsch's term for 'Il n'y a pas de rapport sexuel'. Maybe this brings us to what the 'Lubitsch touch' is at its most elementary: an ingenious way to make this failure work. That is to say, instead of reading the fact that there is no sexual relationship as a traumatic obstacle on account of which a love affair has to end up in some kind of tragic failure, this very obstacle can be turned into a comic resource, it can function as something to be circumvented, alluded to, played with, exploited, manipulated, made fun of – in short, sexualized. Sexuality is an exploit which thrives on its own ultimate failure.

Along these comic lines, one can imagine a different (Lubitschean) ending to Puccini's *Tosca*, where the music remains exactly the same and only the action in the last seconds is changed. Joseph Kerman wrote an acerbic comment on the final notes of *Tosca*, in which the orchestra bombastically recapitulates the 'beautiful' pathetic melodic line of Cavaradossi's 'E lucevan le stelle', as if, unsure of what to do, Puccini simply repeated the most 'effective' melody from the previous score, ignoring all narrative or emotional logic.[9] In Puccini's version, when Cavaradossi is taken by the guards to be executed on the roof of the prison palace, Tosca pulls him aside and explains to him that the execution will be fake, but that he must give a believable performance in order for them to escape freely later. Mario is taken away and Tosca is left waiting impatiently. As the execution is carried out and the guns are fired, Mario falls to the ground. Tosca shouts out, happy with his flawless performance. Once everyone leaves, she rushes to Mario to hug him, overjoyed with the prospect of the new life ahead of them. When she bends down to him, she realizes he is dead – Scarpia has betrayed her from beyond the grave, real bullets were used. Heartbroken, she throws herself over his body and weeps . . . Imagine a simple change: after Tosca realizes that he really is dead and utters a desperate cry, Mario gets up with a laugh and tells her that, by pretending he is dead, he just played a practical joke on her – in reality, Scarpia did keep his word and the bullets were not real. The couple embrace ecstatically to the triumphant music of 'E lucevan le stelle'.

However, as has been noted by perceptive critics, there is a fundamental ambiguity about the key point in *Trouble in Paradise* which is echoed in his *Design for Living* (1933). Lubitsch seems to take the cynical position of respecting appearances while secretly transgressing them. Recall Heinrich Heine's well-known quip that one should value above everything else 'freedom, equality, and crab soup'. 'Crab soup' stands here for all the small pleasures in the absence of which we become (figurative, if not real) terrorists, following an abstract idea and imposing it on reality without any consideration of concrete circumstances. Nowhere is this wise perspective displayed in a clearer way than in *Heaven Can Wait*. At the film's beginning, the old Henry Van Cleve enters the opulent reception area of Hell and is personally greeted by 'His Excellency' (the devil), to whom he tells the story of his dissolute life so that he can secure his place there. After hearing Henry's story, His Excellency denies him entry and suggests he try 'the other place', where his dead wife Martha and his good grandfather are waiting for him – there might be 'a small room vacant in the annex' up there. So the devil is nothing but God himself with a touch of wisdom, not taking small transgressions too seriously, well aware that they make us human. But if the devil is a good, wise man, is then the true Evil not God himself, insofar as he lacks ironic wisdom and blindly insists on obedience to his law? What the devil knows is that there must be some trouble in paradise if we are to enjoy it.

The political stakes are very high here: if benevolent, cynical realism is our horizon - and is this not the whole point of *Ninotchka* (1939), in which pursuit of pleasure wins over ideology? – then a more radical Left should be benevolently mocked. But are things as simple as this? Let's take Jaroslav Hašek, the author of the legendary comical novel *The Good Soldier Švejk* (1923), who is usually perceived as an advocate of the sound common sense of ordinary people against all forms of fanaticism. His novel tells the adventures of an ordinary Czech soldier who undermines the ruling order simply by following orders too literally; Švejk finds himself at the front-line trenches in Galicia, where the Austrian Army is confronting the Russians. When Austrian soldiers start to shoot, the desperate Švejk runs into the no-man's-land area in front of their trenches, desperately waving his hands and shouting: 'Don't shoot! There are men on the other side!' This is what Lenin was aiming at in his call to the tired peasants and other working masses in the summer of 1917 to stop fighting, part of a ruthless strategy to win popular support and thus gain power, even if it meant the military defeat of his own country. The link to Lenin is not so far-fetched if we take into account the much lesser-known fact that, immediately after the First World War, Hašek fought as a *politkommissar* in a Red Army division in the

Russian Civil War. And it's the same with Lubitsch: we should never forget the unexpected moments of the eruption of violent fury in his films, as when the hero of *Trouble in Paradise* explodes to his beloved about the injustice of a system which prosecutes small thieves but tolerates big financial robberies, or the moment in *Cluny Brown* (1946), when the heroine points out (with obvious political implications) that sometimes you can fix things with small corrections, but at others you just have to intervene violently and smash it all to really put it right.

So this brings us back to the beginning: Lubitsch and the Left. I am not trying to make Lubitsch into an undercover Bolshevik, at least not a Stalinist one. As for a Lubitschean approach to Stalinism, we should move beyond *Ninotchka*; the unique figure of Mikhail Suslov[10] would have been a much more appropriate character. Can we, then, imagine a scene, similar to that in *Heaven Can Wait*, in which Lubitsch confronts a severe Bolshevik *kommissar* who has to decide whether Lubitsch will be sent to the gulag or given a post in the Communist Party hierarchy? Knowing how the system works, Lubitsch readily confesses all his petit-bourgeois individualist sins, but the *kommissar* (played by Bela Lugosi, as in *Ninotchka*), just like the benevolent devil in *Heaven Can Wait*, tells him that, unfortunately, the gulag is not the right place for him, and then offers him not a small back room in Heaven but something much more interesting. He remembers one of Lenin's last proposals. We find in late Lenin an unexpected Lubitschean feature: the accent on good manners and humour. Politeness is more than just obeying external legality, and less than pure moral activity – it is the ambiguous, imprecise domain of what one is not strictly obliged to do (if one doesn't do it, one doesn't break any laws), but what one is nonetheless expected to do. We are dealing here with implicit, unspoken regulations, with questions of tact, with something towards which the subject has, as a rule, a non-reflected relationship: something that belongs to our spontaneous sensitivity, a deep seam of customs and expectations which is part of our inherited mores (*Sitten*). Therein resides the self-destructive deadlock of political correctness: it tries to explicitly formulate, legalize even, the stuff of manners – if I look at a woman in a way considered offensive, I not only display bad manners, I violate the law.

Lenin was aware that, no matter how emancipatory the new Bolshevik Master is, he has to be supplemented by another counterbalancing form. As Moshe Lewin noted in his *Lenin's Last Struggle*,[11] at the end of his life Lenin himself intuited this necessity: while fully admitting the dictatorial nature of the Soviet regime, he proposed a new ruling body, the Central Control Commission. What first strikes one is Lenin's unexpected focus on politeness and civility – a strange thing coming from a hardened Bolshevik. Lenin's famous appeal to remove Stalin also concerns the latter's lack of politeness:

> Stalin is too rude, and this defect, though quite tolerable in our midst and in dealings among us Communists, becomes intolerable in a General Secretary. That is why I suggest that the comrades think about a way to remove Stalin from that post and appoint in his place another man who in all respects differs from Comrade Stalin in his superiority, that is, who is more tolerant, more loyal, more courteous and more considerate of the comrades, less capricious, etc.[12]

Although Lenin's struggle against the rule of state bureaucracy is well known, what is less so is that, as Lewin perceptively noted, with his proposal of the Central Control Commission Lenin was trying to square the circle of democracy and dictatorship of the party-state; while fully admitting the dictatorial nature of the Soviet regime, he tried:

> . . . to establish at the summit of the dictatorship a balance between different elements, a system of reciprocal control that could serve the same function – the comparison is no more than approximate – as the separation of powers in a democratic regime. An important Central Committee, raised to the rank of Party Conference, would lay down the broad lines of policy and supervise the whole Party apparatus, while itself participating in the execution of more important tasks . . . Part of this Central Committee, the Central Control Commission, would, in addition to its work within the Central Committee, act as a control of the Central Committee and of its various offshots – the Political Bureau, the Secretariat, the Orgburo. The Central Control Commission . . . would occupy a special position with relation to the

> other institutions; its independence would be assured by its direct link to the Party Congress, without the mediation of the Politburo and its administrative organs or of the Central Committee.[13]

Checks and balances, the division of power, mutual control . . . this was Lenin's desperate answer to the question: who controls the controllers? There is something dreamlike, properly phantasmatic, in this idea of the Central Control Commission: an independent, educational controlling body with an 'apolitical' edge, consisting of the best teachers and technocratic specialists, with neutral expert knowledge, who would keep in check the 'politicized' Central Committee and its organs – in short, the Party executives. However, everything hinges here on the true independence of the Party congress, which is *de facto* already undermined by the prohibition of factions, thus enabling the top Party apparatus to control the congress, dismissing its critics as 'factionalists'. The naivety of Lenin's trust in technocratic experts is all the more striking if we bear in mind that it comes from a politician who was otherwise fully aware that political struggle allows for no neutral position. However, in 'dreaming' (his expression) about the mode of operation of the Central Control Commission, he describes how this body should resort:

> . . . to some semi-humorous trick, cunning device, piece of trickery or something of that sort. I know that in the staid and earnest states of Western Europe such an idea would horrify people and that not a single decent official would even entertain it. I hope, however, that we have not yet become as bureaucratic as all that and that in our midst the discussion of this idea will give rise to nothing more than amusement.
>
> Indeed, why not combine pleasure with utility? Why not resort to some humorous or semi-humorous trick to expose something ridiculous, something harmful, something semi-ridiculous, semi-harmful, etc.?[14]

Is this not an almost obscene double of the 'serious' executive power concentrated in the Central Committee and Politburo, a kind of *non-organic intellectual* of the movement – an agent resorting to humour, tricks and the cunning of reason, keeping itself at a distance – a kind of *analyst*? So, maybe, could we not imagine Lubitsch presiding over this Control Commission? The obvious counter-argument here is: but did the authoritarian structure of Bolshevik power not preclude a figure such as Lubitsch from playing any key role? In reply we should quote Hegel's comment, from the 'Introduction' to his *Phenomenology*, that 'the standard for the examination is altered when that for which it is supposed to be the standard itself fails the examination, and the examination is not merely an examination of knowledge but also that of the standard of knowledge'. Brutally applied to our case, this means that, if Lubitsch doesn't fit Lenin, then we need a new Lenin, a Lenin who would tolerate, demand even, a figure like Lubitsch at the head of his Control Commission.

A Leninist Gesture in *LA LA LAND* AND IN *BLACK PANTHER*

We find traces of this other 'Leninist' Lubitsch, not just the cynically benevolent Lubitsch, in many other later Hollywood products – one just has to know how and where to look for them. Let's take Damien Chazelle's *La La Land* (2016); among the politically correct criticisms of it, the one that stands out for its sheer stupidity is that there are no gay couples in the film, which takes place in LA, a city with a strong gay population. How come those politically correct Leftists who complain about the under-representation of sexual and ethnic minorities in Hollywood movies never complain about the gross misrepresentation of the lower-class majority of workers – it's OK if workers are invisible, as long as here or there a gay or lesbian character appears? I remember a similar incident at the first Conference on the Idea of Communism in London in 2009. Some members of the public voiced the complaint that there was only one woman among the participants, plus no black person and no one from Asia; to which Alain Badiou remarked that it was strange how no one was bothered by the fact that there were no workers among the participants even though the topic was Communism.

Returning to *La La Land*, we should bear in mind that the movie opens with the depiction of hundreds of precarious and/or unemployed workers on their way to Hollywood to search for a job that would boost their career. The first song ('Another Day of Sun') shows them singing and dancing to pass the time while they are stuck in a highway traffic jam. Mia and Sebastian, who are among them in their separate cars, are the two who will make the big time, the obvious exceptions. And, from this standpoint, their falling in love (which will enable their success) enters the story to blur in a background of invisibility the hundreds who will not make it, the implication being that it is their love (and not sheer luck) which makes them special and destined to succeed.[15] Ruthless competition is the name of the game, with no hint of solidarity (recall the numerous audition scenes where Mia is repeatedly humiliated). No wonder that, when I hear the first lines of the film's most famous song ('City of stars, are you shining just for me? City of stars, there's so much that I can't see'), I find it hard to resist the temptation to hum back the most stupid orthodox Marxist reply imaginable: 'No, I am not shining just for a petit-bourgeois ambitious individual like you, I am also shining for the thousands of exploited, precarious workers in Hollywood whom you can't see and who will not succeed like you, to give them some hope!'

Mia and Sebastian start a relationship and move in together, but they grow apart because of their desire to succeed: Mia's ambition is to become an actress while Sebastian wants to own a club where they would play authentic old jazz. First Sebastian joins a pop-jazz band and spends time touring, then, after the premiere of her monodrama fails, Mia leaves Los Angeles and moves back home to Boulder City. Alone in LA, Sebastian receives a call from a casting director who had attended and enjoyed Mia's play, and invites Mia to a film audition. Sebastian drives to Boulder City and persuades her to return. At the audition Mia is asked to tell a story; she begins to sing about her aunt, who inspired her to pursue acting. Confident that the audition was a success, Sebastian asserts that Mia must devote herself wholeheartedly to the opportunity. They profess they will always love each other, but are uncertain of their future. Five years later, Mia is a famous actress and married to another man, with whom she has a daughter. One night the couple stumble upon a jazz bar. Noticing the 'Seb's' logo, Mia realizes that Sebastian has finally opened his own club. Sebastian spots Mia, looking unsettled and regretful, in the crowd and begins to play their love theme; this prompts an extended dream sequence in which the two imagine what might have been had their relationship worked out perfectly. The song ends and Mia leaves with her husband. Before walking out, she shares with Sebastian one last knowing look and smile, happy for the dreams they have both fulfilled.

As was noted by many critics, the final ten-minute fantasy shows how the story would have been told in a classic Hollywood musical. Such a reading confirms the film's reflexivity: it portrays how the movie should end with regard to the genre formula to which it relates. *La La Land* is clearly a self-reflexive film, a film about the genre of musicals, but it works alone, and one doesn't have to know the full history of musicals to enjoy and understand it. As André Bazin wrote of Chaplin's *Limelight* (1952): it is a reflexive film about the old Chaplin's declining career, but it stands alone, one doesn't have to know about Chaplin's early career as the Tramp to enjoy it. Interestingly, the more we progress into *La La Land*, the fewer musical numbers there are and the more it becomes pure melodrama – until, at the end, we are thrown back into a musical which explodes as a fantasy.

Apart from obvious references to other musicals, Chazelle nods subtly to Mark Sandrich's classic Rogers/Astaire musical screwball comedy, *Top Hat* (1935). There are many good things to say about *Top Hat*, beginning with the role of tap-dancing as a disturbing intrusion into the daily routine (Astaire practises tap-dancing on the hotel floor above Ginger Rogers, prompting her to complain, which brings the couple together); we should also mention the topic of marriage retroactively declared invalid; the process of false marriage and its repetition; a rich man's servant is mad at his master for wearing a wrong tie and refuses to talk to him . . . Compared to *La La Land*, what cannot but strike us is the total psychological flatness of *Top Hat*, where there is no depth, just puppet-like acting which pervades even the most intimate moments. The final song and its staging ('Piccolino') in no way relates to the story's happy ending; the words of the song are purely self-referential, they merely tell the story of how this song itself came to be and invite us to dance to it:

By the Adriatic waters Venetian sons and daughters
Are strumming a new tune upon their guitars.
It was written by a Latin, a gondolier who sat in
His home out in Brooklyn and gazed at the stars.
He sent his melody across the sea to Italy,
And we know they wrote some words to fit that catchy bit
And christened it the Piccolino.
And we know that it's the reason why
Everyone this season is strumming and humming a
 new melody.
Come to the casino and hear them play the Piccolino.
Dance with your bambino to the strains of the catchy
 Piccolino.
Drink your glass of vino, and when you've had your
 plate of scalopino,
Make them play the Piccolino, the catchy Piccolino,
And dance to the strains of that new melody, the Piccolino.

And this is the truth of the film: not the ridiculous plot, but the music and tap-dancing as goals in themselves. There is a parallel with Hans Christian Andersen's *The Red Shoes*: the heroine just cannot help dancing, it is for her an irresistible drive. The singing dialogue between Astaire and Rogers, even at its most sensuous (as in the famous 'Dancing Cheek to Cheek'), is just a pretext for music and dancing. *La La Land* may rank as superior, since it dwells in the realm of psychological realism: reality intrudes into the dream world of musicals (like the latest instalments of superhero films, which bring out the hero's psychological complexity, his traumas and inner doubts). But it is crucial to note how the otherwise realistic story has to conclude with an escape into musical fantasy.

The first and obvious Lacanian reading of *La La Land* would see its plot as yet another variation on the theme of 'there is no sexual relationship': the successful careers of Mia and Sebastian, which tear them apart, are like the *Titanic* hitting an iceberg in James Cameron's movie. Their careers are there to preserve the dream of love (staged in the final fantasy), that is, to mask the inherent impossibility of their love, the fact that, if they were to remain together, they would turn into a bitter, disappointed couple. Consequently, the ultimate version of the film would have been the reversal of the final situation: Mia and Sebastian are together and enjoy full professional success, but their lives are empty, so they go to a club and dream of a fantasy in which they happily live a modest life with each other, having both renounced their careers.

We encounter a similar reversal in *The Family Man* (Brett Ratner, 2000). Jack Campbell, a single Wall Street executive, hears on Christmas Eve that his former girlfriend, Kate, has called him after many years. On Christmas Day, Jack wakes up in a suburban New Jersey bedroom with Kate and two children; he hurries back to his office and condo in New York, but his closest friends do not recognize him. He is now living the life he could have had, had he stayed with his girlfriend (a modest family life, where he is a car-tyre salesman for Kate's father and Kate is a non-profit lawyer). Just as Jack finally realizes the true value of his new life, his epiphany jolts him back to his wealthy former life. He forgoes closing a big acquisition deal in order to intercept Kate, who has also focused on her career and become a wealthy corporate lawyer. After he learns that she merely called him to give back some of his old possessions since she is moving to Paris, he runs after her at the airport and, in an effort to win back her love, describes the family they had in the alternate universe. She agrees to have a cup of coffee at the airport, suggesting that they might have a future . . . So what we get is a compromise solution at its worst: somehow the two will combine the best of both worlds, remaining rich capitalists but being at the same time a loving couple with humanitarian concerns – in short, they will have their cake and eat it.

La La Land at least avoids this cheap optimism. So what, effectively, happens at the film's end? It's not, of course, that Mia and Sebastian simply decide to give preference to their careers over their relationship. At the very least one should add that they both find success and achieve their dreams precisely because of the relationship they had, so that their love is a kind of vanishing mediator: far from being an obstacle to their success, it 'mediates' it. So does the film subvert the Hollywood image of a couple – Mia and Sebastian both fulfil their dreams, but *not* as a couple? And is this subversion more than simply a postmodern narcissistic preference of personal fulfilment over love? In other words, what if their love was not a true love-Event? Plus, what if their career 'dream' was not devotion to a true artistic Cause but just that, a dream? So what if none of the competing claims (career, love) really display an 'evental' dimension, an unconditional commitment that follows a true Event? Their love is not true, their pursuit of their careers is not a full artistic commitment. In short, Mia's and Sebastian's betrayal goes deeper than making a choice which compels them to renounce the alternative: their entire life is already a betrayal of an authentically committed existence. This is also why the tension between the two claims (of love and career) is not a tragic existential dilemma but a very soft uncertainty.[16]

Such a reading is nonetheless too simple – it ignores the enigma of the final fantasy: *whose* fantasy is it, his or hers? Is it not hers: she is the observer-dreamer, and the whole dream is focused on her destiny – she goes to Paris to shoot the film, etc.? Against some critics who claim that the film is male-biased, i.e. that Sebastian is the active partner in the couple, one should assert that Mia is the subjective centre-point of the film: the choice is much more hers than his, which is why, at the end, she is the big star and Sebastian, far from being a celebrity, is just the owner of a moderately successful jazz club (which also sells fried chicken). This difference becomes clear when we listen closely to the two conversations between Mia and Sebastian in which one of them has to make a choice. When Sebastian announces to her that he will join the band and spend most of the time touring, Mia does not raise the question of what this means for the two of them; instead, she asks him if this is what *he* really wants and enjoys, and Sebastian replies that the public like what he is doing, so his playing with the band means a permanent job and a career, with the opportunity to put some money aside and open his own jazz club. But she insists, correctly, that the true question is that of his own desires: what bothers her is not that, if he chooses his career (playing with the band), he will betray her (their love relationship), but that, if he chooses this career, he will betray *himself*, his true vocation. In the second conversation, which takes place after the audition, there is no conflict and no tension: Sebastian immediately recognizes that for Mia acting is not just a career opportunity but a true vocation, something she has to do to be herself; abandoning it would ruin the very basis of her personality, and he implores her to do it without any reserve or regret. There is no choice here between their love and her vocation: in a paradoxical but deeply true sense, if she were to abandon the prospect of acting in order to stay with him in LA, she would also betray their love, since it grew out of their shared commitment to a Cause.

We stumble here upon a problem ignored by Badiou in his theory of Event: if the same subject is addressed by multiple Events, which of them should be given priority? What should an artist decide to do if he or she cannot combine his/her love life (building a life together with a partner) with his/her dedication to art? We should reject the very terms of this choice: in an authentic dilemma, one should not decide between Cause and love, between fidelity to one or the other. The authentic relationship between Cause and love is more paradoxical. The basic lesson of King Vidor's *Rhapsody* (1954) is that, in order to gain his beloved's love, the man has to prove that he is able to survive without her, that he prefers his mission or profession to her. There are two immediate choices: (1) my professional career is what matters most to me, the woman is just an amusement, a distracting affair; (2) the woman is everything to me, I am ready to humiliate myself, to forsake all my public and professional dignity for her. They are both false, they lead to the man being rejected by the woman. The message of true love is thus even if you are everything to me, I can survive without you, I am ready to forsake you for my mission or profession. The proper way for the woman to test the man's love is thus to 'betray' him at the crucial moment of his career (the first public concert in the film, the key exam, the business negotiation which will decide his career) – only if he can survive the ordeal and successfully accomplish his task although deeply traumatized by her desertion, will he deserve her and she will return to him. The underlying paradox is that love, as the Absolute, should not be posited as a direct goal; it should retain the status of a by-product, of something we receive as an undeserved grace. Perhaps there is no greater love than that of a revolutionary couple, where each of the two lovers is ready to abandon the other at any moment if the revolution demands it.

The question is, how does an emancipatory-revolutionary collective which embodies the 'general will' affect intense erotic passion? From what we know about love among the Bolshevik revolutionaries, something unique took place, a new form of love emerged: a couple living in a permanent state of emergency, totally dedicated to the revolutionary Cause, ready to sacrifice all personal sexual fulfilment to it, even prepared to abandon and betray each other if revolution demanded it, but simultaneously totally dedicated to each other, enjoying rare moments together with extreme intensity. The lovers' passion was tolerated, even silently respected, but ignored in the public discourse as something of no concern to others. (There are traces of this even in what we know of Lenin's affair with Inessa Armand.) There is no attempt at *Gleichschaltung*, at enforcing the unity between intimate passion and social life: the radical *disjunction* between sexual passion and social-revolutionary activity is fully recognized. The two dimensions are accepted as totally heterogeneous, each irreducible to the other, there is no harmony between the two – but it is this very recognition of the gap which makes their relationship non-antagonistic.

And does the same not happen in *La La Land*? Does Mia not make the 'Leninist' choice of her Cause, does Sebastian not support her in it, and do they in this way not remain faithful to their love?

Although such a reading, of course, goes against the grain of the predominant perception of the film, it can be justified by the simple fact that it offers the only way to read consistently some important details of the film's texture. The same holds for *Black Panther*, an otherwise very ambiguous product whose final moments nonetheless point to another type of Leninist fidelity.[17] The first sign of the film's ambiguity is the fact that it was enthusiastically received all across the political spectrum: from partisans of black emancipation, who saw in it the first big Hollywood assertion of black power, through modest Left-liberals, who sympathized with its reasonable solution – education and help, not struggle – up to some representatives of the alt-right, who easily recognized in the film's affirmation of black identity and way of life another version of Trump's 'America first' (and, incidentally, this is why Mugabe, before he lost power, also said some kind words about Trump). When all sides recognize themselves in the same product, we can be sure that the product in question is ideology at its purest, i.e. a kind of empty vessel containing antagonistic elements.

Here is the plot:[18] centuries ago, five African tribes fight over a meteorite containing vibranium. One warrior ingests a 'heart-shaped herb' affected by the metal and gains superhuman abilities, becoming the first 'Black Panther'; he unites all but the Jabari tribe to form the nation of Wakanda. The Wakandans use the vibranium to develop advanced technology and isolate themselves from the world by posing as an undeveloped Third World country.[19] – In 1992, King T'Chaka visits his undercover brother N'Jobu in Oakland, California. T'Chaka accuses N'Jobu of assisting black-market arms-dealer Ulysses Klaue in stealing vibranium from Wakanda. N'Jobu's partner reveals he is Zuri, another undercover Wakandan, and confirms T'Chaka's suspicions.

In the present day, following T'Chaka's death, his son T'Challa returns to Wakanda to assume the throne. He and Okoye, the leader of the Dora Milaje regiment (a feminine praetorian guard of the court), extract Nakia, T'Challa's ex-lover, from an undercover assignment so she can attend his coronation ceremony with his mother Ramonda and younger sister Shuri. At the ceremony, the Jabari Tribe's leader M'Baku challenges T'Challa for the crown in ritual combat. T'Challa defeats M'Baku and convinces him to yield rather than die.

Meanwhile, Klaue and Erik Stevens, a young black militant, steal a Wakandan artifact from a London museum; W'Kabi, T'Challa's friend and Okoye's lover, urges him to bring Klaue back dead or alive. T'Challa, Okoye, and Nakia travel to Busan (South Korea) where, in a casino, Klaue plans to sell the artifact to CIA agent Everett K. Ross. A firefight erupts and Klaue attempts to flee but is caught by T'Challa who reluctantly releases him to Ross's custody. Klaue tells Ross that Wakanda's international image is a front for a technologically advanced civilization. Erik attacks and extracts Klaue as Ross is severely injured protecting Nakia. Rather than pursue Klaue, T'Challa takes Ross to Wakanda where their technology can save him.

While Shuri heals Ross, T'Challa confronts Zuri about N'Jobu. Zuri explains that N'Jobu planned to share Wakanda's technology with people of African descent around the world to help them conquer their oppressors. As T'Chaka arrested N'Jobu, N'Jobu attacked Zuri, forcing T'Chaka to kill him. T'Chaka ordered Zuri to lie that N'Jobu had disappeared and left behind N'Jobu's American son, Erik, who became a US black ops soldier specialized versed in the art of subverting Third World governments which pose a threat to the US interests. Meanwhile, Erik (who adopted the name 'Killmonger') kills Klaue and takes his body to Wakanda. He is brought before the tribal elders, revealing his identity and claim to the throne. Killmonger challenges T'Challa to ritual combat; after killing Zuri, he defeats T'Challa and hurls him over a waterfall. After ingesting the heart-shaped herb, Killmonger orders the rest incinerated, but Nakia extracts one first. Killmonger, supported by W'Kabi and his army, prepares to distribute shipments of Wakandan weapons to operatives around the world.

Nakia, Shuri, Ramonda and Ross flee to the Jabari Tribe for aid. They find a comatose T'Challa, rescued by the Jabari in repayment for sparing M'Baku's life. Healed by Nakia's herb, T'Challa returns to fight Killmonger, who dons his own Black Panther suit and commands W'Kabi and his army to attack T'Challa. Shuri, Nakia, and the Dora Milaje join T'Challa, while, acting on Shuri's instructions, Ross remotely pilots a jet and shoots down the planes carrying the vibranium weapons to the underground revolutionaries around the globe. M'Baku and the Jabari arrive to reinforce T'Challa, and, confronted by Okoye, W'Kabi and his army stand down. Fighting in Wakanda's vibranium mine, T'Challa disrupts Killmonger's suit and stabs him. Killmonger refuses to be healed, choosing to die a free man rather than love as a *de facto* slave.

In the final scene, T'Challa establishes an outreach centre at the building where N'Jobu died; the centre, to be run by Nakia and Shuri, will distribute knowledge and help to the underprivileged. In an additional mid-final-credits scene, T'Challa, Okoye, and Nakia are visiting the United Nations, and T'Challa tells the assembled delegation that he wants to share Wakanda's advancements with the world, but in a peaceful way, through supporting education and rendering available its superior technology.

Already the starting point seems problematic or ambiguous, at least: recent history teaches us that being blessed by some precious natural resource is rather a curse in disguise – think about today's Congo, which is an inoperative 'rogue state' with drugged child soldiers and so on precisely because of its incredible wealth of natural resources (and the way they are ruthlessly exploited for it). The scene then shifts to Oakland, which was one of the strongholds of the real Black Panthers, a radical black liberation movement from the 1960s that was ruthlessly suppressed by the FBI (members were killed, etc.). Following the path of the Black Panther comics, the movie – without ever directly mentioning the real Black Panthers – in a simple but no less masterful stroke of ideological manipulation efficiently kidnaps the name, so that its first association is now no longer the old radical militant organization but a superhero-king of a powerful African kingdom. More precisely, there are two Black Panthers in the film, the king T'Challa and his cousin Erik, and each of them stands for a different political vision. Erik spent his youth in Oakland and then as a US army black-op, and his domain is poverty, gang violence and military brutality, while T'Challa was raised in the secluded opulence of the Wakanda's royal court; consequently, Erik advocates a militant global solidarity (Wakanda should put its wealth, knowledge, and power at the disposal of the oppressed all around the world so that they could overthrow the existing world order), while T'Challa is slowly moving from the traditional isolationism of 'Wakanda first!' to a gradual and peaceful globalism which would act within the coordinates of the existing world order and its institutions, spreading education and technological help, and simultaneously maintaining the unique Wakanda culture and way of life. (This is why T'Challa is more a doubtful hero torn between different paths of action than the usual hyper-active superhero, while his opponent Killmonger is always ready to act and knows what to do.) The fact that a CIA agent plays a key role in his final victory tells it all . . . (One can argue that the task to destroy the planes sent by Killmonger to supply weapons to revolutionaries around the world is given to the white CIA agent because the king doesn't trust his own black compatriots – they may sabotage the mission because they sympathize with Killmonger.)

While many critics praised the active role of women in the Wakanda court, plus the wealth of their different positions (defense, old-age wisdom, science and technology . . .), such an assertion of femininity is strictly subordinate to male domination. So even with Wakanda's opening towards the world, all that will change is that a dosage of traditional wisdom will contain the excesses of wild capitalism. With T'Challa steering the helm, today's rulers can continue to sleep in peace.

One of the signs that something is wrong with this picture is the strange role of the two white characters, the 'bad' South-African Klaue and the 'good' CIA agent Ross. The 'bad' Klaue doesn't fit the role of the villain for which he is predestined – he is too weak and comical. Ross is a much more enigmatic figure, in some sense the symptom of the film: a CIA agent – and this means that his superiors, and thereby the US government, know the truth about Wakanda – who participates in the events with an ironic distance, strangely non-engaged, as if he is participating in a show. Why is he selected by Shuri as the pilot of the plane that will shoot down the planes carrying the weapons to Killmonger's agents all around the world (the role reserved for 'good' blacks in the usual sci-fi Armageddon movies)? Isn't it that he holds the place of the existing global system in the film's universe? And at the same time holding the place of us, the majority of white viewers of the film, as if telling us: 'It's OK to enjoy this fantasy of black supremacy, none of us is really threatened by this alternate universe!'

This leaves Killmonger as the only true villain – a revolutionary Black Panther from 1960s versus the mythic royal Black Panther from our comic superhero universe . . . The fact that T'Challa opens up to 'good' globalization but is also supported by its repressive embodiment, the CIA, demonstrates that there is no real tension between the two: 'return to roots' fits perfectly with global capitalism, which can only be undermined through a different global project. So we should not be fascinated by the beautiful spectacle of Wakanda's capital as an alternate modern city where technology serves human needs, where tradition and ultra-modernity seamlessly blend together: what this beautiful spectacle obliterates is the insight followed by Malcolm X when he adopted the name X. The point of choosing X as his family name and thereby signaling that the slave traders who brought the enslaved Africans from their homeland brutally deprived them of their family and ethnic roots, of their entire cultural life-world, was not to mobilize the blacks to fight for the return to some primordial

African roots, but precisely to seize the opening provided by X, an unknown new (lack of) identity engendered by the very process of slavery which made the African roots for ever lost. The idea is that this X which deprives the blacks of their particular tradition offers a unique chance to redefine (reinvent) themselves, to freely form a new identity much more universal than white people's professed universality. (As is well known, Malcolm X found this new identity in the universalism of Islam.) This precious lesson of Malcolm X is forgotten by *Black Panther*: to attain true universality, a hero must go through the experience of losing his/her roots.

Things thus seem clear, confirming Fred Jameson's insistence on how difficult it is to imagine a really new world, a world which does not just reflect (invert, supplement) the existing world order. One can only wonder at how a critic could have actually written that Frantz Fanon, the theorist of black liberation through violent rebellion, would have been 'proud of the movie' . . . However, there are signs that disturb this simple and obvious reading, signs that push us towards reading the film in the way Leo Strauss read Plato's and Spinoza's work, as well as Milton's *Paradise Lost*. Although he used terms like 'secret teaching', Strauss was not a Gnostic engaged in a 'deep hermeneutics'; he was not looking for the esoteric Plato in the sense of some hidden knowledge to be decoded from the public text. Everything is shown and said, all alternative theories are clearly presented; a careful Straussian reading just draws attention to signs that indicate that the obvious hierarchy of theoretical positions has to be inverted.

For example, Book I of Plato's *Republic* deals with the polemical dialogue between Socrates and Thrasymachus, who violently disagrees with the outcome of Socrates' discussion with Polemarchus about justice: Thrasymachus claims that 'justice is the advantage of the stronger' (338c) and that 'injustice, if it is on a large enough scale, is stronger, freer, and more masterly than justice' (344c). Socrates counters by forcing him to admit that there is a standard of justice beyond the advantage of the stronger. However, a close reading makes it clear what Plato's real position is: with regard to facts, Thrasymachus is right, justice *is* the advantage of the stronger – but this truth should be kept secret since its public disclosure would hurt and demoralize the majority of ordinary people whose moral sensitivity demands that right is stronger than might. It's the same with Milton's *Paradise Lost*: although he follows the church's official line of condemning Satan's rebellion, Milton's sympathies are clearly with Satan. (We should add that it doesn't matter if this preference for the 'bad' side or agent is conscious or unconscious to the author of a text, the result is the same.) Does the same not hold for Christopher Nolan's *Dark Knight Rises*, the final part of his Batman trilogy? Although Bane is the official villain, there are indications that he, much more than Batman himself, is the film's authentic hero, distorted as its villain: he is ready to sacrifice his life for his love, ready to risk everything for what he perceives as injustice, and this basic fact is occluded by superficial and rather ridiculous signs of destructive evil.

So, back to *Black Panther*: what are the signs that enable us to recognize Killmonger as its true hero? There are many, the first being the scene of his death, where the heavily wounded Erik prefers to die free than to be healed and survive in the false abundance of Wakanda – the strong ethical impact of Killmonger's last words immediately ruin the idea that he is a simple villain. What then follows is a scene of extraordinary warmth: the dying Killmonger sits down at the edge of a mountain precipice observing the beautiful Wakanda sunset, and his cousin T'Challa who has just defeated him silently sits at his side. There is no hatred here, just two basically good men with a different political view sharing their last moments after the battle is over – a scene unimaginable in a standard action movie, which culminates in the vicious destruction of the enemy.[20] These final moments alone cast a doubt on the film's obvious reading and solicit from us a much deeper reflection.

Conclusion: For How Long Can We Act Globally and Think Locally?

What can we learn from Hegel about Donald Trump and his liberal critics? Quite a lot, surprisingly. In his critical account of Romantic irony, Hegel scathingly dismisses it as an exercise of empty negativity, of the vain subjectivity which perceives itself as elevated over every objective content, making fun of everything, caught in 'the hither and thither course of the humour which uses every topic only to emphasize the subjective wit of the author':

> it is the artist himself who enters the material, with the result that his chief activity, by the power of subjective notions, flashes of thought, striking modes of interpretation, consists in destroying and dissolving everything that proposes to make itself objective and win a firm shape for itself in reality, or that seems to have such a shape already in the external world. Therefore every independence of an objective *content* along with the inherently fixed connection of the *form* (given as that is by the subject matter) is annihilated in itself, and the presentation is only a sporting with the topics, a derangement and perversion of the material, and a rambling to and fro, a criss-cross movement of subjective expressions, views, and attitudes whereby the author sacrifices himself and his topics alike.[1]

Hegel's point is usually taken as conservative: instead of the all-destroying anarchic irony of the Romantics, one should recognize the Good and True embodied in social customs, i.e. its own rational core. However, Hegel is much more ambiguous here. First, his basic reproach to subjective humour is not that it undermines all objective content, not taking it seriously, relativizing it, but that this all-destroying ironic stance is really utterly impotent: it actually threatens nothing, it just provides the ironic subject with the illusion of inner freedom and superiority. When individuals are caught in an impenetrable cobweb of social relations, the only way to assert their subjectivity is with the cache of jokes which allegedly demonstrate their inner superiority.

Hegel contrasts with Romantic subjective irony a much more radical ontological irony which characterizes the innermost core of dialectics: apropos Socratic irony, he points out that, 'like all dialectic, it gives force to what is – taken immediately, but only in order to allow the dissolution inherent in it to come to pass; and we may call this the universal irony of the world'.[2] The dialectical approach does not try actively to undermine reality, which it perceives as an antagonist; it just lets it be what it is (or claims to be), taking it more seriously than it takes itself, and in this way allows it to destroy itself. This irony is in a way objective, so no wonder that, in a short (and regrettably underdeveloped) passage, Hegel contrasts subjective humour with what he calls 'objective humour':

> what matters to humour is the object and its configuration within its subjective reflex, then we acquire thereby a growing intimacy with the object, a sort of objective humor ... The form meant here displays itself only when to talk of the object is not just to name it, not an inscription or epigraph which merely says in general terms what the object is, but only when there are added a deep feeling, a felicitous witticism, an ingenious reflection, and an intelligent movement of imagination which vivify and expand the smallest detail through the way that poetry treats it.[3]

We are dealing here with a humour which, by way of focusing on significant symptomal details, brings out the inherent inconsistencies/antagonisms of the existing order. So would it not be legitimate to extrapolate from these indications the idea that social totality itself is traversed by antagonisms, wrought by comical reversals? Freedom turns into terror, honour into flattery – are such reversals not the stuff of the Cunning of Reason? Can one imagine a more terrifying case of objective humour than that of Stalinism, of the comical reversal of great emancipatory hopes into self-destructive terrorist violence? Was, in this sense, Stalin not the big jokester of the twentieth century? And is, in our time, individual freedom of choice also not a joke, the truth behind which is the desperate situation of the precarious worker? In view of the fact that the greatest cultural product of the Stalinist era was political jokes, one is tempted yet again to paraphrase Brecht: what is even the best anti-Stalinist joke compared to the joke that is Stalinist politics itself? Or, closer to our time, what are even the best jokes about Trump compared to the joke that is Trump's actual politics? Imagine that, a couple of years ago, a comedian performed on stage Trump's statements, tweets and decisions – it would have been experienced as a non-realist, exaggerated joke. So Trump is already his own parody, with the uncanny effect that the reality of his behaviour is more outrageously funny than most parodies of it.

Insofar as objective humour displays the inner negativity of the thing itself, no wonder many interpreters of Hegel, from Dieter Henrich on, see in objective humour the last resort of modern art, of art after the end of art. One can even venture a step further here and propose the concept of absolute humour, the unity of subjective and objective humour: the insight into how objective humour needs subjective humour in order to reproduce itself. In this sense, Hegel's critique of subjective humour is today more actual than ever. One of the popular myths of the late-Communist regimes in Eastern Europe was that there was a department of the secret police whose function was not to collect but to invent and circulate political jokes against the regime and its representatives, as they were aware of jokes' positive stabilizing function: political jokes offer ordinary people an easy and tolerable way to blow off steam, easing their frustrations.

And, on a different level, the same holds for Trump. Remember how many times the liberal media announced that Trump was caught with his pants down and had committed public suicide (mocking the parents of a dead war hero, boasting about pussy-grabbing, etc.). Arrogant liberal commentators were shocked at how their continuous acerbic attacks on Trump's vulgar racist and sexist outbursts, factual inaccuracies, economic nonsense and so on did not hurt him at all but maybe even enhanced his popular appeal. They missed how identification works: we as a rule identify with others' weaknesses, not only, or even principally, with their strengths; so the more Trump's limitations were mocked, the more ordinary people identified with him and perceived attacks on him as condescending attacks on themselves. To them, the subliminal message of Trump's vulgarities was: 'I am one of you!', and they felt constantly humiliated by the liberal elite's supercilious attitude towards them. As Alenka Zupančič succinctly put it, 'the extremely poor do the fighting for the extremely rich, as was clear in the election of Trump. And the Left does little else than scold and insult them.'[4] Or, we should add, the Left does what is even worse: it patronizingly 'understands' the confusion and blindness of the poor. This Left-liberal arrogance explodes in its purest form in the new genre of political-comment-comedy talk shows (Jon Stewart, John Oliver), which mostly showcase the pure sense of superiority of the liberal intellectual elite:

> Parodying Trump is at best a distraction from his real politics; at worst it converts the whole of politics into a gag. The process has nothing to do with the performers or the writers or their choices. Trump built his candidacy on performing as a comic heel – that has been his pop culture persona for decades. It is simply not possible to parody effectively a man who is a conscious self-parody, and who has become president of the United States on the basis of that performance.[5]

In my past work, I used a joke from the good old days of 'really existing socialism' which was popular among dissidents. In fifteenth-century Russia, which was occupied by Mongols, a farmer and his wife walk along a dusty country road. A Mongol warrior on a horse stops at their side and tells the farmer that he will now rape his wife; he then adds: 'But since there is a lot of dust on the ground, you should hold my testicles while I'm raping your wife, so that they will not get dirty!' After the Mongol finishes the job and rides away, the farmer starts to laugh and jump for joy. The surprised wife asks him: 'How can you be jumping for joy when I was just brutally raped in your presence?' The farmer answers: 'But I got him! His balls are full of dust!' This sad joke tells of the predicament of dissidents: they thought they were dealing serious blows to the party *nomenklatura*, but all they were doing was getting a little bit of dust on the *nomenklatura*'s testicles, while the *nomenklatura* went on raping the people. And can we not say exactly the same about the likes of Jon Stewart making fun of Trump – do they not just dust his balls, and in the worst cases scratch them?

The basic lesson of dialectics is thus that vain subjective humour should be countered not by serious 'objective' analysis but by Hegel's objective humour, which brings out the absurdities inherent in our reality. We should not be afraid to discern this humorous aspect even in our most terrifying experiences. After it opened in Berlin in 2015, Ferdinand von Schirach's *Terror* became a global hit, with hundreds of performances all around the world, as well as an unending flow of ethical debates in the mass media. It is a court drama, the account of the trial against Lars Koch, a German fighter pilot who has shot down a Lufthansa plane that has been hijacked by a terrorist; the plane was heading for a stadium of 70,000 people (watching a Germany/England game), and Koch's pragmatic decision – one in which he broke constitutional law – was to end the lives of 164 people on the plane rather than allow the terrorist to slaughter a far greater number in the stadium. At the end, the audience must vote on his guilt: each spectator is provided with a small gadget with two buttons, 1 (guilty) or 2 (not guilty), and the audience learns its verdict. Predictably, the majority, at least in Western theatres, proclaim Koch not guilty.

We are undoubtedly dealing with a genuine antinomy of moral reason (to use a Kantian turn) here: if we formulate the dilemma in this clear way, there simply is no unambiguous solution. Any play with certainty and numbers – in the style of, 'If I am absolutely sure that by killing one man I will save at least fifty, then . . .' – amounts to an obscenity. Our gut feeling that there is something deeply wrong and false with the choice posed by the play is fully justified: the choice is ideology at its purest, mainly because of what it leaves out in order to present a clear and simple picture. Basically, we are addressed as individuals, confronted with a tough choice whose very clarity (shoot down the plane or not?) obfuscates all other relevant questions. What about emptying the stadium (there was enough time), what about the geopolitical causes of such terrorist acts, what about our military interventions into Arab countries, what about our alliance with Saudi Arabia? Did we choose any of that, were we asked to choose any of it? Why do we only feel the pressure of the choice when we are confronted with a consequence of all these previous choices?

But there is another, more basic, aspect of the play that we should address. Upon a closer look, we realize that, when Koch chooses to shoot down the plane, he does not really make an individual existential decision but just follows the implicit social injunction. His conversations with his military superiors make it clear that they assume he will shoot down the plane; they even subtly put pressure on him to do it, they just don't want to tell him to do it straight out. The situation again recalls the story about the ashtray at the beginning of this book, where a contradiction between prohibition and permission is openly assumed and thereby cancelled out, treated as non-existent – the message was: 'It's prohibited, and here it is how you do it.' Was Koch's situation not exactly the same? The repeated message from his superiors was: 'It's prohibited by the law . . . and do it!'

This is how armies function. I remember a similar incident from my military service. One morning, the first class was on international military law, and among other rules the officer mentioned that it was prohibited to shoot at parachuters while they are still in the air, before they touch the ground. In a happy coincidence, our next class was about rifle-shooting, and the same officer taught us how to target a parachuter in the air (how, while aiming at him, one should take into account the velocity of his descent and the direction and strength of the wind, and so on). When one of the soldiers asked the officer about the contradiction between this lesson and what we had learned just an hour before, the officer just snapped back with a cynical laugh: 'How can you be so stupid? Don't you understand how life works?'

We should note here that, in the case of nuclear war, the popular imagination yearns for the opposite scenario: that of a single officer who resists the command to press the button and thus saves the world. Recall a frightening detail from the Cuban Missile Crisis: only later did we learn how close to nuclear war we were during a naval skirmish between an American destroyer and a Soviet B-59 submarine off Cuba on 27 October 1962. The destroyer dropped depth charges near the submarine to try to force it to surface, not knowing it had a nuclear-tipped torpedo. Vadim Orlov, a member of the submarine crew, told the conference in Havana that the submarine was authorized to fire it if three officers agreed. The officers began a fierce, heated debate over whether to sink the ship. Two of them said yes and the other said no. 'A guy named Arkhipov saved the world,' was the bitter comment of a historian on this incident.[6] Do we not all silently count on something similar happening in the fiery exchange between the US, North Korea and others – that, at a decisive moment, a single individual will find the strength to stop the mad cycle of nuclear threats and counter-threats?

We can imagine a series of choices along the lines of von Schirach's play: if a North Korean missile on its way to Guam falls apart, for example, should the US respond, and how? But what we should always bear in mind is the madness of the entire situation: while we are all threatened by ecological catastrophes, we continue to play games of self-destruction. The decisions our leaders consider are not on the scale of 'How many innocent people am I allowed to kill to save many more?', but 'How many millions of innocent bystanders I am ready to kill, directly and indirectly, in order to strike back at the enemy?' This is what they are really talking about when they evoke the catastrophic consequences of a nuclear conflict: millions and millions will die, but, somehow, we will have to do it and strike back.

What further complicates matters is that, if one listens to Kim Jong-un when he talks about dealing a devastating blow to the US, one cannot but wonder how he sees his own position. He talks as if he is not aware that his country, himself included, will be destroyed; it is as if he is playing a fantasy game. So is he bluffing, is he not really considering a nuclear strike? If the basic underlying axiom of the Cold War was that of MAD (Mutually Assured Destruction), that of today's nuclear games seems to be the opposite, that of NUTS (Nuclear Utilization Target Selection) – the idea that, by means of a surgical strike, we can destroy the enemy's nuclear capacities while the anti-missile shield is protecting us from a counter-strike. More precisely, the US adopts a different strategy: it acts as if it continues to trust the MAD logic in its relations with Russia and China, while at the same time being tempted to practice NUTS with Iran and North Korea. The paradoxical mechanism of MAD inverts the logic of the 'self-realizing prophecy' into a 'self-stultifying intention': the very fact that each side can be sure that, if it decides to launch a nuclear attack on the other, the other side will respond with full destructive force, guarantees that no side will start a war. The logic of

NUTS is, on the contrary, that the enemy can be forced to disarm if it knows that we can strike at it without risking a counter-attack. The very fact that two directly contradictory strategies are adopted simultaneously by the same superpower bears witness to the phantasmatic character of this entire reasoning. The only thing we can do in such a situation is to mobilize the broadest international public in order to directly criminalize *any* talk of the use of nuclear and other weapons of mass destruction. Leaders and states who even consider it should be treated as pariahs, as obscene subhuman monsters. Anything should be permitted against them, from mass boycott to personal humiliation.

The looming military conflict between the US and North Korea contains a double danger. Although both sides are for sure bluffing, not counting on an actual nuclear exchange, rhetoric never functions as mere rhetoric but can always run out of control. Furthermore, as many commentators have noticed, the weird thing is that Trump decided to occupy a position symmetrical to that of Kim Jong-un, raising the stakes in the game. This escalation increasingly resembles the struggle for recognition between the two subjects described by Hegel, in which the winner is the one who proves his readiness to die rather than make a compromise on behalf of life. Trump thereby inadvertently got caught up in a game which does not become a true superpower – in the case of a small and weak country like North Korea, a discreet, stern warning would be enough. To act otherwise is ridiculous.

With the first steps of a historic appeasement between North and South Korea from the early 2018, the tone magically changed into one of mutual respect and collaboration, and what causes unease is precisely the swiftness of this change, as if we are part of a strange game in which sudden reversals prevent us taking any of it seriously. But are such reversals also possible between Israel and Iran, between the US and Russia? Crazy rumours even began to circulate about Trump deserving a Nobel Peace prize – will he get it? The French have a beautiful expression, *Voyons voir*, which can be roughly translated as 'Let's wait and see what we'll see.' Four US Presidents have previously won a Nobel Peace prize: Theodore Roosevelt, Woodrow Wilson, Jimmy Carter (after leaving office) and Barack Obama, in

2009, for his 'extraordinary efforts to strengthen international diplomacy and cooperation between peoples' – an explanation which was a fake, merely expressing the hope that Obama would act like that in the future.

Unbelievable as the proposal for Trump to get the Nobel Peace prize is, we should nonetheless do three things in reacting to it. First, we should bear in mind that the great compromise which enabled the breakthrough towards a peaceful resolution of the Korean crisis was made not by Trump but by Kim Jong-un: Kim made the key concession, so the prize should be delivered to Kim and Trump, and the ridicule of this idea is obvious – a Peace prize for the head of arguably the most oppressive regime in the world? Should Donald and Kim really be awarded just for performing a sudden U-turn and not acting as crazily as we feared? Furthermore, how to combine the Peace-prize-for-Trump initiative with Trump's belligerent withdrawal from the pact with Iran? The fact that his withdrawal was opposed by all Western European allies of the US opens up new possibilities for a global geopolitical rearrangement: it could isolate the US from the community of states which will continue to adhere to the pact, thereby further reducing the US to just one among the global powers.

However, the unpleasant truth (for Left liberals) is that, far from being just the bellicose crazy US leader, Trump doesn't turn out so bad in comparison with Hillary Clinton. Asked by the *Guardian* whether she truly believes Clinton would be more dangerous than Trump, Susan Sarandon responded:

> I did think she was very, very dangerous. We would still be fracking, we would be at war [if she was president]. It wouldn't be much smoother. Look what happened under Obama that we didn't notice. She would've done it the way Obama did it, which was sneakily. He deported more people than have been deported now. How he got the Nobel Peace prize I don't know.[7]

We should thus always bear in mind that, at his worst, Trump is mostly just continuing the politics of his predecessors.

When, after announcing the meeting with Kim Jong-un, Trump decided to withdraw from the Iran agreement, it may have appeared that we saw two Trumps in quick succession, the 'peaceful' Trump, whose acts led to the prospect of disarmament in Korea, and the 'belligerent' Trump, who decided to withdraw from the Iran pact and thereby brought instability and the threat of war (not only) to the Middle East. But there is only one Trump who was doing exactly the same in both cases. In the case of North Korea, he began with exerting extreme pressure, inclusive of economic sanctions and military threat, and he is doing the same to Iran, maybe with the hope that, if it worked the first time, it will work now also . . . Will it? We should not forget the obvious difference between the two cases: North Korea is an isolated state with no interests in the wider region, while Iran is deeply involved in the Middle East conflicts. One of the few events that raised a small glimmer of hope in this dangerous mess was the surprising direct meeting between the leaders of South Korea and North Korea towards the end of May 2018, after Trump cancelled the meeting with Kim. Perhaps, this is the only proper way to proceed: to bypass the US meddling and release tensions at the local level, without the 'help' of the superpowers.

But what if the US is well aware that the pressure on Iran will not work? What if, together with Israel and Saudi Arabia, it is preparing for war with Iran? It is difficult to speculate about the consequences of such a military conflict, especially if we bear in mind that, in the long term, it is not possible to prevent a country acquiring nuclear (or chemical or biological) weapons. However, what is not difficult to see is the absurdity of the very notion of Trump as the candidate for the Nobel Peace prize – who, then, really deserves it? Those who, for sure, will never get it. Recall Sophia Karpai, the head of the cardiographic unit of the Kremlin Hospital in the late 1940s. Her (accidental) misfortune was that it was her job to twice take an electrocardiogram (ECG) of Andrei Zhdanov, the Communist leader, on 25 July 1948 and 31 July, days before Zhdanov's death through heart failure. The first ECG, taken after Zhdanov displayed some heart troubles, was inconclusive (a heart attack could be neither confirmed nor excluded), while the second one surprisingly showed a much better picture (the intraventricular blockage had disappeared, a clear indication that there was no heart attack). In 1951, she was arrested on the charge that, in conspiracy with other doctors treating Zhdanov, she had falsified the data, erasing the clear indications that a heart attack *had* occurred, thereby depriving Zhdanov of the special care needed by a victim of heart attack. After harsh treatment, including continuous brutal beating, all other accused doctors confessed. 'Sophia Karpai, whom Vinogradov [her boss] had described as nothing more than "a typical person of the street with the morals of the petty bourgeoisie", was kept in a refrigerated cell without sleep to compel a confession. She did not confess.'[8]

The impact and significance of her perseverance cannot be overestimated: her signature would have dotted the i on the prosecutor's case on the 'doctor's plot', immediately setting in motion the mechanism that, once rolling, would lead to the death of hundreds of thousands, maybe even to a new European war (according to Stalin's plan, the 'doctor's plot' should have demonstrated that the Western intelligence agencies tried to murder the top Soviet leaders, and thus served as an excuse to attack Western Europe). She persisted just long enough for Stalin to enter his final coma, after which the entire case was immediately dismissed. Her simple heroism was crucial in the series of details which, 'like grains of sand in the gears of the huge machine that had been set in motion, prevented another catastrophe in Soviet society and politics generally, and saved the lives of thousands, if not millions, of innocent people.'[9]

This simple persistence against all odds is ultimately the stuff true heroes are made of. We learn about such cases only sometimes and only years later. So, if there is to be a minimal justice in who gets the Nobel Peace prize, this prize should be given neither to active politicians for their present acts (i.e., for just not being as brutal as one expected them to be) nor to politicians for their future expected acts; the prize should be given retroactively, to nameless heroes like Arkhipov and Karpai.

Back to our main line, the logic of nation-state competition is extremely dangerous because it runs directly against the urgent need to establish a new way of relating to our environs, a radical politico-economic change called by Peter Sloterdijk 'the domestication of the wild animal culture'. Until now, each culture disciplined and educated its own population and guaranteed civic peace among them in the guise of state power; but the relationship between different cultures and states was permanently under the shadow of potential war, with each period of peace being nothing more than a temporary armistice. As Hegel conceptualized it, the entire ethic of a state culminates in the highest act of heroism, the readiness to sacrifice one's life for one's nation state, which means that the wild, barbarian relations between states serve as the foundation for the ethical life within them. Is today's North Korea, with it ruthless pursuit of nuclear weapons and rockets with which to hit distant targets, not the ultimate example of this logic of unconditional nation-state sovereignty? However, the moment we fully accept the fact that we live on Spaceship Earth, the task that urgently imposes itself is that of civilizing civilizations themselves, of imposing universal solidarity and co-operation among all human communities – a task rendered all the more difficult by the ongoing rise of sectarian religious and ethnic 'heroic' violence and readiness to sacrifice oneself (and the world) for one's specific Cause.

Addressing members of the Russian parliament, Vladimir Putin said on 1 March 2018: 'The missile's test-launch and ground trials make it possible to create a brand-new weapon, a strategic nuclear missile powered by a nuclear engine. The range is unlimited. It can manoeuvre for an unlimited period of time. No one in the world has anything similar.' To applause he concluded: 'Russia still has the greatest nuclear potential in the world, but nobody listened to us. Listen to us now.'[10]

Yes, we should listen to these words, but we should listen to them as to the words of a madman joining the duet of two other madmen. Remember how, a little while ago, Kim Jong-un and Donald Trump competed about buttons to trigger nuclear missiles that they have at their disposal, with Trump claiming his button is bigger than Kim's. Now we have Putin joining this obscene competition – which is, we should never forget, a competition about who can destroy us all more quickly and efficiently – with the claim that his is the biggest . . . Each side can, of course, claim that it wants only peace and is only reacting to the threat posed by others (for example, Putin immediately added that he is just reacting to Trump's claims that, due to its protective shields, the US can win nuclear war with Russia) – true, but what this means is that the madness is in the whole system itself, in the vicious cycle we are caught in once we participate in the system. The structure is here similar to that of the supposed belief where also all individual participants act rationally, attributing irrationality to the other who reasons in exactly the same way. From my youth in Socialist Yugoslavia, I remember a weird incident with toilet paper. All of a sudden, a rumour started to circulate that there was not enough toilet paper in the stores. The authorities promptly issued assurances that there was enough toilet paper for normal use, and, surprisingly, this was not only true but people mostly believed it was true. However, an average consumer might reason in the following way: I know there is enough toilet paper and the rumour is false, but what if some people take this rumour seriously and, in a panic, start to buy excessive reserves of toilet paper, causing in this way an actual shortage? So I'd better go and buy lots of it myself. It is not even necessary to believe that some others will take the rumour seriously – it is enough to presuppose that some others believe, like you, that there will be people who take the rumour seriously; the effect is the same, namely a real lack of toilet paper in the stores. (Is the phenomenon of bitcoins not just this structure – an entity whose only substance is that subjects believe in it – brought to extreme? The mantra that, at some point, bitcoin will have to collapse, because it is based on nothing, expresses our fear – not the fear of this collapse but the opposite fear that something based on nothing can nonetheless persist indefinitely.)

This madness becomes visible the moment we ask a simple question: how do the protagonists in the nuclear standoff (Kim, Trump, Putin) imagine pressing the button? Are they not aware of the almost 100 per cent certainty that their own country will also be destroyed by retaliatory strikes? They are aware . . . and also not, i.e. they clearly speak from a schizophrenic position: although they know they will perish, they talk as if they somehow stand outside the danger and can strike at the enemy from a safe place. This schizophrenic position combines the two axioms of nuclear warfare, MAD and NUTS. The very fact that two directly contradictory strategies are mobilized simultaneously by the same superpower bears witness to the fantasmatic character of this entire reasoning. In December 2016, this inconsistency reached an almost unimaginably ridiculous peak: both Trump and Putin emphasized the chance for new more friendly relations between Russia and the US, and simultaneously asserted their full commitment to the arms race – as if peace among the superpowers can only be provided by a new Cold War. Alain Badiou has written that the contours of the future war are already drawn:

> the United States and their Western–Japanese clique on the one side, China and Russia on the other side, atomic arms everywhere. We cannot but recall Lenin's statement: 'Either revolution will prevent the war or the war will trigger revolution.' This is how we can define the maximal ambition of the political work to come: for the first time in History, the first hypothesis – revolution will prevent the war – should realize itself, and not the second one – a war will trigger revolution. It is effectively the second hypothesis which materialized itself in Russia in the context of the First World War, and in China in the context of the Second. But at what price! And with what long-term consequences![11]

There is no way to avoid the conclusion that a radical social change – a revolution – is needed to civilize our civilizations. We cannot afford to hope that a new war will lead to a new revolution: a new war would much more probably mean the end of civilization as we know it, with the survivors (if any) organized in small authoritarian groups. However, the main obstacle to this process of civilizing civilizations is not so much sectarian fundamentalist violence as its apparent opposite, cynical indifference. In October 2017, Donald Trump declared a public-health emergency in response to what he called the 'national shame and human tragedy': the US's escalating opioid epidemic, the 'worst drug crisis in American history', caused by the mass prescription of opioid painkillers: 'The United States is by far the largest consumer of these drugs, using more opioid pills per person than any other country by far. No part of our society – not young or old, rich or poor, urban or rural – has been spared this plague of drug addiction.' Although Trump is as far as one can imagine from being a Marxist, his proclamation cannot but evoke Marx's well-known characterization, in his 'Contribution to the Critique of Hegel's Philosophy of Right', of religion as the 'opium of the people', which is worth quoting here:

> Religion is the sigh of the oppressed creature, the heart of a heartless world, and the soul of soulless conditions. It is the opium of the people. The abolition of religion as the illusory happiness of the people is the demand for their real happiness. To call on them to give up their illusions about their condition is to call on them to give up a condition that requires illusions. The criticism of religion is, therefore, in embryo, the criticism of that vale of tears of which religion is the halo.

One immediately notices that Trump, who wants to begin his war on opioids by prohibiting the most dangerous drugs, is a very vulgar Marxist, similar to those hardline Communists (like Enver Hoxha or the Khmer Rouge) who tried to undermine religion by simply outlawing it. Marx's approach is more subtle: instead of directly fighting religion, the goal of Communists is to change the social situation of exploitation and domination which gives birth to the need for religion in the first place. Marx nonetheless remains all too naive, not only in his idea of religion but with regard to the different forms which the opium of the people takes. It is true that radical Islam is an exemplary case of religion as the opium of the people: a false confrontation with capitalist modernity which allows Muslims to dwell in their ideological dream while their countries are ravaged by the effects of global capitalism – and exactly the same holds for Christian fundamentalism. In January 2018, it was reported that lawmakers in Egypt

> are seriously considering passing a law that would make atheism illegal. Blasphemy is already illegal in Egypt, and people are frequently arrested for insulting or defaming religion under the country's strict laws. The newly proposed rule would make it illegal for people not to believe in God, even if they don't talk about it.[12]

Two questions immediately arise here: how will authorities establish if someone is an atheist if he doesn't even talk about it? An ironic comment: will they scan the suspect's brain with the devices used by neuro-theologists trying to determine if there are traces of religious experiences in his neurons? Second question: how do they justify this extreme measure? Here is what Khaled Salah, in his article 'The Atheists are Coming', argues:

> The dangers of terrorism are known, but not many know that atheism and terror are equally destructive. Atheism, also, weakens one's identity and calls into questions established beliefs in history, canons, religious symbols, the Prophet's companions and followers, and ultimately leads to the collapse of the foundations of entire nations and of their sacred beliefs.[13]

So it's not religious fundamentalism but atheism which is to blame for terror, even if it is done in the name of religion! This line of argumentation brings to mind the reaction of the American Catholic Church to the wave of paedophilia among its priests: they evoked some dubious research that put the blame on the secular-hedonist culture that infected the priests . . . The sad thing is that, since, in the last decade, atheism with an Islamophobic twist has established itself as a respectable choice in the US public space (atheism of the Harris/Pinker/Hitchens kind, of course), it has become a fashion in some 'radical' Leftists circles to downplay the critique of religion since it may 'serve the enemy' . . . However, there are today, in our Western world, two other forms of the opium of the people: the opium and the people. As the rise of populism demonstrates, the opium of the people is also 'the people' itself, the fuzzy populist dream destined to obfuscate our own antagonisms. And, last but not least, for many among us the opium of the people is opium itself, escape into drugs – precisely the phenomenon Trump is talking about.

As always, to produce (not only literal but also ideological) opioids – like 'the people' – one needs a very sophisticated technological apparatus. If there is a figure who stands out as the hero of our time, it is Christopher Wylie, a gay Canadian vegan who, at 24, came up with an idea that led to the foundation of a company called Cambridge Analytica, a data analytics firm that went on to claim a major role in the Leave campaign during Britain's EU membership referendum; later, he became a key figure in digital operations during Donald Trump's election campaign, creating Steve Bannon's psychological warfare tool. Wylie's plan was to break into Facebook and harvest the Facebook profiles of millions of people in the US, and to use their private and personal information to create sophisticated psychological and political profiles, and then target them with political ads designed to work on their particular psychological makeup. At a certain point, Wylie was genuinely freaked out: 'It's insane. The company has created psychological profiles of 230 million Americans. And now they want to work with the Pentagon? It's like Nixon on steroids.'[14]

What makes this story so fascinating is that it combines elements which we usually perceive as opposites. The 'alt-right' presents itself as a movement that addresses the concerns of ordinary, white, hardworking, deeply religious people who stand for simple traditional values and abhor corrupted eccentrics like homosexuals and vegans, and also digital nerds – and now we learn that their electoral triumphs were masterminded and orchestrated precisely by such a nerd who stand for all they oppose . . . There is more than an anecdotal value in this fact: it clearly signals the vacuity of alt-right populism, which has to rely on the latest technological advances to maintain its popular redneck appeal. Plus it dispels the illusion that being a marginal computer nerd automatically stands for a 'progressive' anti-system position.

So where does this need to escape into opium come from? To paraphrase Freud, we have to take a look at the psychopathology of global-capitalist everyday life. Yet another form of today's opium of the people is our escape into the pseudo-social digital universe of Facebook, Twitter, and so on. In a speech to Harvard graduates in May 2017, Mark Zuckerberg told his public: 'Our job is to create a sense of purpose!' – and this from a man who, with Facebook, has created the world's most expanded instrument of purposeless waste of time!

As Laurent de Sutter demonstrated, chemistry in its scientific form is becoming part of us: large aspects of our lives are characterized by the management of our emotions by drugs, from everyday use of sleeping pills and anti-depressants to hard narcotics. We are not just controlled by impenetrable social powers, our very emotions are 'outsourced' to chemical stimulation. The stakes of this chemical intervention are double and contradictory: we use drugs to keep external excitement (shocks, anxieties, and so on) under control, to desensitize us to them, and to generate artificial excitement if we are depressed and lack desire. Drugs are thus deployed against the two opposed threats to our daily lives, over-excitement and depression, and it is crucial to notice how these two uses of drugs relate to our private and public life: in the developed Western countries, our public lives increasingly lack collective excitement (for example, that provided by genuine political engagement), while drugs supplant this lack with private (or, rather, intimate) forms of excitement – they euthanize public life and artificially excite private life.[15] (What remains of the passionate public engagement in the West is mostly the populist hatred, and this brings us to the other second opium of the people, the people itself.) The country which is most impregnated with this tension is South Korea; here is Franco Berardi's report on his recent journey to Seoul:

> Korea is the ground zero of the world, a blueprint for the future of the planet . . . After colonization and wars, after dictatorship and starvation, the South Korean mind, liberated by the burden of the natural body, smoothly entered the digital sphere with a lower degree of cultural resistance that virtually any other populations in the world. In the emptied cultural space, the Korean experience is marked by an extreme degree of individualization and simultaneously it is headed towards the ultimate cabling of the collective mind. These lonely monads walk in the urban space in tender continuous interaction with the pictures, tweets, games coming out of their small screens, perfectly insulated and perfectly wired into the smooth interface of the flow . . . South Korea has the highest suicide rate in the world. Suicide is the most common cause of death for those under 40 in South Korea. Interestingly, the toll of suicides in South Korea has doubled during the last decade . . . in the space of two generations their condition has certainly improved by the point of view of revenue, nutrition, freedom and possibility of travelling abroad. But the price of this improvement has been the desertification of daily life, the hyper-acceleration of rhythms, the extreme individualization of biographies, and work precariousness which also means unbridled competition . . . The intensification of the rhythm of work, the desertification of the landscape and the virtualization of the emotional life are converging to create a level of loneliness and despair that is difficult to consciously refuse and oppose.[16]

What Berardi's impressions of Seoul provide is the image of a place deprived of its history, a worldless place. Badiou noted that we live in a social space which is progressively experienced as worldless. Even Nazi anti-Semitism, however ghastly it was, opened up a world: it described its critical situation by positing an enemy which was a 'Jewish conspiracy'; it named a goal and the means of achieving it. Nazism disclosed reality in a way which allowed its subjects to acquire a global 'cognitive mapping', which included a space for their meaningful engagement. Perhaps it is here that one should locate one of the main dangers of capitalism: although it is global, it sustains a *sensu stricto* worldless ideological constellation, depriving the large majority of people of any meaningful cognitive mapping. This, then, is what makes millions of us seek refuge in our opiums: not just new poverty and lack of prospects, but unbearable superego pressure in its two aspects – the pressure to succeed professionally and the pressure to enjoy life fully in all its intensity. Perhaps this second aspect is even more unsettling: what remains of our life when our retreat into private pleasure itself becomes the stuff of brutal injunction? In short, is Trump himself – the way he acts, emitting endless tweets, etc. – not the cause of the disease he is trying to cure?

Back in the 1960s, the motto of the early ecological movements was, 'Think globally, act locally!' With his politics of sovereignty mirroring the stance of North Korea, Trump promises to do the exact opposite, to turn the US into a global power, but this time in the sense of, 'Act globally, think locally!' We should not be afraid to add that this locus has a precise name – we think locally because we are caught in a Plato's cave of ideology – so how do we get out of it? Here we encounter an intricate dialectics of freedom and servitude:

> The exit from the cave begins when one of the prisoners is not only freed from his chains (as Heidegger shows, this is not at all enough to liberate him from the libidinal attachment to the shadows), but when he is forced out. This clearly must be the place for the (libidinal, but also epistemological, political and ontological) function of the master. This can only be a master who neither tells me precisely what to do nor whose instrument I could become, but must be one who just 'gives me

back to myself'. And in a sense, one might say this could be connected to Plato's anamnesis theory (remembering what one never knew, as it were) and does imply that the proper master just affirms or makes it possible for me to affirm that 'I can do this', without telling me.[17]

The point Ruda makes here is a subtle one: it's not only that, if I am left by myself in the cave, even without chains, I prefer to stay there, so that a master has to force me out – I have to volunteer to be forced out, similarly to the way in which, when a subject enters psychoanalysis, he or she does it willingly, voluntarily accepting the psychoanalyst as his or her master (albeit in a very specific way):

> A question arises at precisely this point from using the reference to the master in psychoanalytic terms: does this mean that those who need a master are – always – in the position of the analysand? If – politically – such a master is needed to become who one is (to use Nietzsche's formula), and this can be structurally linked to liberating the prisoner from the cave (to forcing him out after the chains are taken off and he still does not want to leave), the question arises how to link this with the idea that the analysand must constitutively be a *volunteer* (and not simply slave or bondsman). So, in short, there must be a dialectics of master and volunteer(s): a dialectics because the master to some extent constitutes the volunteers as volunteers (liberates them from a previously seemingly unquestionable position), such that then they become voluntary followers of the master's injunction, whereby the master ultimately becomes superfluous – but maybe only for a certain period of time; afterwards one has to repeat this very process (one never succeeds in leaving the cave entirely, so that one constantly has to re-encounter the master (and the anxiety linked to it), i.e., a master's intervention is needed if things get stuck again, or mortifyingly habitualized).

What further complicates the picture is that:

> capitalism relies massively on unpaid and thereby structurally 'voluntary' labour. There are, to put it in Lenin's terms, volunteers and 'volunteers', so maybe one has to distinguish not only between different types of master figures but also link them (if the link to psychoanalysis is in this way pertinent) to different understandings of the

volunteer, the analysand. Even the analysand, a volunteer, must somehow be forced into analysis. This might seem to bring classical readings of the master-slave dialectics back onto the stage, but I think one should bear in mind that as soon as the slave identifies himself as a slave he is no longer a slave, whereas the voluntary worker in capitalism can identify himself as what he is and this changes nothing (capitalism interpellates people as 'nothings', volunteers, etc.).

The two levels of volunteering (which are simultaneously two levels of *servitude volontaire*) are different, not only with regard to the context of servitude (to market mechanisms, to an emancipatory cause); their very form is different. In capitalist servitude we simply feel free, while in authentic liberation we accept voluntary servitude as serving a Cause and not just ourselves. In today's cynical functioning of capitalism, I know very well what I am doing and continue to do it, the liberating aspect of my knowledge is suspended, while in the authentic dialectics of liberation my awareness of my situation is already the first step towards liberation. In capitalism I am enslaved precisely when I 'feel free': this feeling is the very form of my servitude, while in an emancipatory process I am free when I 'feel like a slave' – that is to say, the feeling of being enslaved already bears witness to the fact that, in the core of my subjectivity, I am free; only when my position of enunciation is that of a free subject can I experience my servitude as an abomination. Thus we have here two versions of the Möbius strip reversal: if we follow capitalist freedom to the end it turns into the very form of servitude, and if we want to break away from capitalist *servitude volontaire* our assertion of freedom again has to assume the form of its opposite, of voluntarily serving a Cause.

If Marx defined bourgeois human rights as those of '*liberté-égalité-fraternité* and Bentham', the proletarian and properly Leftist version should be 'liberty-equality-freedom and . . . *terror*', the terror of being torn out of the complacency of bourgeois life and its egotist struggles, terror as the pressure to elevate ourselves to universal emancipation. Bentham or terror – this, perhaps, is our ultimate choice. Why is terror needed? Because our chains in the cave today are not those of traditional ideology. Robert Pippin recently pointed out this shift:

> the complexity of our situation has created something quite unprecedented that only [Hegel's] philosophy, with its ability to explain the 'positive' role of the negative, and the reality of group agency and collective subjectivity, can account for. Life in modern societies seems to have created the need for uniquely dissociated collective doxastic states, a repetition of the various characters in the drama of self-deceit narrated by the *Phenomenology*. This is one wherein we sincerely believe ourselves committed to fundamental principles and maxims we are actually in no real sense committed to, given what we do . . . The principles can be consciously and sincerely acknowledged and avowed, but, given the principles they are, cannot be integrated into a livable, coherent form of life. The social conditions for self-deceit in this sort of context can help show that the problem is not rightly described as one where many individuals happen to fall into self-deceit. The analysis is not a moral one, not focused on individuals. It has to be understood as a matter of historical *Geist*.[18]

The key point in this passage is Pippin's emphasis on the 'positive' role of the negative, and 'the reality of group agency and collective subjectivity': the 'negative', in this case, is the dissonance, the gap between explicit ideological texture and its actual practice in the real world. Its positive role means that this dissonance does not prevent the full implementation of an ideology but makes it 'livable', is a condition of its actual functioning – if we take away the negative side, the ideological edifice itself falls apart. The emphasis on 'group agency and collective subjectivity' means that we are not merely concerned with the imperfections of individuals; in that case, the guilt would be that of each person, with his or her corruption and moral depravity, and the cure would be sought in their moral improvement. What we are dealing with is a dissonance inscribed into the 'objective' social spirit itself, into the basic structure of social customs. Such collective forms of self-deceit function as ways of objective social being, and are thus in some sense 'true' even if they are false.

In her *America Day by Day* (1948), Simone de Beauvoir noted: 'many racists, ignoring the rigors of science, insist on declaring that even if the psychological reasons haven't been established, the fact is that blacks *are* inferior. You only have to travel through America to be convinced of it.'[19] Her point about racism has often been misunderstood: for example, Stella Sandford claims that 'nothing justifies Beauvoir's [. . .] acceptance of the "fact" of this inferiority'; 'With her existentialist philosophical framework, we might rather have expected Beauvoir to talk about the *interpretation of* existing physiological differences in terms of inferiority and superiority [. . .] or to point out the mistake involved in the use of the value judgements "inferior" and "superior" to name alleged properties of human beings, as if to "confirm a given fact".'[20] It is clear what bothers Sandford here. She is aware that Beauvoir's claim about the inferiority of blacks aims at something more than the simple fact that, in the American South of the time (and later), blacks were treated as inferior by the white majority and, in a way, effectively *were* inferior. But her critical solution, careful to avoid racist claims about the factual inferiority of blacks, is to relativize their inferiority as being the interpretation and judgement of white racists, and distance it from the question of their very being. But what this softening distinction misses is the truly trenchant dimension of racism: the 'being' of blacks (as of whites or anyone else) is socio-symbolic. When they are treated as inferior by whites, this does indeed make them inferior at the level of their socio-symbolic identity. In other words, white racist ideology exerts a performative efficiency: it doesn't merely interpret what blacks are, it determines their very being and social existence.

We can now see why Sandford and other critics of Beauvoir resist her formulation that blacks actually *were* inferior: this resistance is itself ideological, based on the fear that, if we concede this point, we will have lost the inner freedom, autonomy and dignity of the human individual. Which is why such critics insist that blacks are not inferior but are merely 'inferiorized' by the violence of white racist discourse – an imposition which does not, however, affect them in the very core of their being, and which they can and do resist as free, autonomous agents in their acts, dreams and projects.

Pippin is right to point out that Hegel's description of such collective self-deceit is much more relevant to our times than his positive institutional solutions. There is, however, one problem with his diagnosis: it accords with Hegel's 'progressive' dialectics, where unearthing inconsistency leads to self-cancellation, while in actual life the dissonance of an ideology is its ultimate stability: only in a specific situation – a change in ideological sensitivity – does the realization that our ideological edifice is dissonant lead to its disintegration. For example, although slavery was obviously incompatible with Christian morality, it took a long time before it became intolerable for the majority.

Pope Francis usually displays the right intuitions in matters theological and political. Recently, however, he committed a serious blunder in endorsing the proposal, propagated by some Catholics, to change the line in the Lord's Prayer which asks God to 'lead us not into temptation': 'It is not a good translation because it speaks of a God who induces temptation. I am the one who falls; it's not him pushing me into temptation to then see how I have fallen. A father doesn't do that, a father helps you to get up immediately. It's Satan who leads us into temptation, that's his department.' So the pontiff suggests we should all follow the Catholic Church in France, which already uses the phrase 'do not let us fall into temptation' instead.[21] Convincing as this simple line of reasoning may sound, it misses the deepest paradox of Christianity and ethics. Is God not exposing us to temptation in paradise, where he warns Adam and Eve not to eat the apple from the tree of knowledge – why did he put this tree there in the first place, and then even draw attention to it? Was he not aware that human ethics can arise only after the Fall? Many perceptive theologians and Christian writers, from Kierkegaard to Paul Claudel, were fully aware that, at its most basic, temptation arises in the form of the Good – or, as Kierkegaard put it apropos Abraham, when he is ordered to slaughter Isaac, his predicament 'is an ordeal such that, please note, the ethical is the temptation'.[22] Is the temptation of the false Good not what characterizes all forms of religious fundamentalism?

Here is a perhaps surprising historical example: the killing of Reinhard Heydrich. The Czechoslovak government-in-exile in London resolved to assassinate him; Jan Kubiš and Jozef Gabčík, who headed the team chosen for the operation, were parachuted in the vicinity of Prague. On 27 May 1942, alone with his chauffeur, Klein, in an open car to show his courage and trust, Heydrich was on his way to his office; when, at a junction in a Prague suburb, the car slowed, Gabčík stepped in front of it and took aim with a submachine gun which, however, jammed. Instead of ordering his driver to speed away, Heydrich decided to halt the car and confront his attackers. At this moment, Kubiš threw a bomb at the rear of the car, and the explosion wounded both himself and Heydrich. When the smoke cleared, Heydrich emerged from the wreckage with his gun in his hand; he chased Kubiš for half a block but grew weak from shock and collapsed. He sent Klein to chase Gabčík on foot, while, pistol still in hand, he gripped his left flank, which was bleeding profusely. A Czech woman went to his aid and flagged down a delivery van; he was first placed in the driver's cab, but complained that the van's movement was causing him pain, so he was put in the back, on his stomach, and quickly taken to the emergency room at a nearby hospital. (Incidentally, although Heydrich died a couple of days later, there was a serious chance that he would have survived, so this woman could well have entered history as the person who saved his life.) While a militarist Nazi sympathizer would emphasize Heydrich's personal courage, what fascinates me is the role of the anonymous Czech woman: she helped Heydrich, who was lying alone in blood with no military or police protection. Was she aware who he was? If yes, and if she was no Nazi sympathizer (both the most probable surmises), why did she do this? Was it the simple, instinctive reaction of human compassion, of helping a neighbour in distress no matter who he or she (or ze, as we will be soon forced to add) is? Should this compassion trump the awareness of the fact that this 'neighbour' was a top Nazi criminal responsible for thousands (and later millions) of deaths? What we are confronted with here is the ultimate choice between abstract liberal humanism and the ethics implied by radical emancipatory struggle: if we follow liberal humanism to its logical conclusion, we find ourselves condoning the worst criminals; and if we do the same with partial political engagement, we end up on the side of emancipatory universality – in which case, the poor Czech woman would have resisted her compassion and tried to finish Heydrich off.

Such impasses are the stuff of actual, engaged ethical life, and if we exclude them as problematic we are left with a lifeless, benevolent holy text. What lurks behind this exclusion is the trauma of the Book of Job, where God and Satan directly organize the destruction of Job's life in order to test his devotion. Quite a few Christians claim the Book of Job should therefore be excluded from the Bible as pagan blasphemy. However, before we succumb to such politically correct ethic cleansing, we should pause for a moment to consider what we lose with it. If we want to keep the Christian experience alive, let us resist the temptation to purge from it all 'problematic' passages, which are the very stuff which confers on Christianity the unbearable tensions of true life.

And the same goes for the viability of a state. As Fred Jameson perceptively noted, Sophocles' *Antigone* is not the story of the disintegration of the organic unity of mores (*Sittlichkeit*), of the division of this unity into public and private (family) customs. Rather, the ethical conflict *Antigone* describes is constitutive of public order; it is a story about the constitution, not disintegration, of state power. Because of this limitation, Pippin too seems to miss the full extent of today's self-deceit when he describes it in quantitative terms ('an even more widespread phenomenon', etc.):

> But collective self-deceit of the kind explored by Hegel is [today] a different and arguably an even more widespread phenomenon ... 'Political actors are presented, and present themselves,' [Bernard] Williams suggests, 'like actors in a soap opera, playing roles in which they neither cynically pretend to represent positions they know to be false (not always or mostly, anyway), nor, given the theatricality, exaggeration, "posing," and the "protest too much" rhetoric, do they comfortably and authentically inhabit those roles.' Williams's description is memorable. 'They are called by their first names or have the same kind of jokey nicknames as soap opera characters, the same broadly sketched personalities, the same dispositions to triumph and humiliations which are schematically related to the doings of other characters. One believes in them as one believes in characters in a soap opera: one accepts the invitation to half believe them.' He goes on to say that 'politicians, the media, and the audience conspire to pretend that important realities are being considered, that the actual word is being responsibly addressed. And of course it is not being addressed. The whole strategy is an attempt to avoid doing so.' ... this is all best accounted for by saying that *Geist*, in this case, the

> communal *Geist* of a nation, is, in its self-representations, engaging in collective self-deceit . . . this is exactly the situation we find ourselves in, in anonymous mass societies, in which the absence of what, according to Hegel, amounts to genuine commonality, *Sittlichkeit*, is a felt absence, not merely an indeterminate absence.[23]

However, the fact that in our 'anonymous mass societies' the absence of *Sittlichkeit* 'is a felt absence, not merely an indeterminate absence' in no way precludes the possibility that *Sittlichkeit* works here as a retroactive dream obscuring the fact that its own reality implies dissonance. Furthermore, do the quoted passages from Williams which describe political actors as characters in a soap opera, although beautifully written, really deliver what they promise to? Do they really describe a new form of moral corruption? Is the fact that 'politicians, the media, and the audience conspire to pretend that important realities are being considered, that the actual word is being responsibly addressed' not a feature of every ideology in its actual functioning? In every ideology, the clear division between the deceived and their deceivers is blurred, since the deceived comply with the illusion and even want to be deceived. What is happening today is not just more of the same, but a qualitatively new form of dissonance: one openly admitted, and for that reason treated as irrelevant. The paradox is thus that, today, there is in a sense less deception than in the way ideology functioned in the past: nobody is really deceived.

In other words, it is not that prior to our current era we took the rules and prohibitions seriously while today we openly violate them. What changed are the rules which regulate appearances, i.e. what can appear in public space. Let's compare the sexual lives of two US presidents, Kennedy and Trump. As we know now, Kennedy had numerous affairs, but the press and TV ignored all this, while Trump's every (old and new) step is followed by the media – not to mention that Trump also speaks publicly in an obscene way that we cannot even imagine Kennedy doing. The gap that separates the dignified public space from its obscene underside is now more and more transposed into public space, with ambiguous consequences: inconsistencies and violations of public rules and openly accepted or at least ignored, but, simultaneously, we are all becoming openly aware of these inconsistencies.

At this point we reach the supreme irony: as it functions today, ideology appears as its exact opposite, as a radical critique of ideological utopias. The predominant ideology now is not a positive vision of some utopian future but a cynical resignation, an acceptance of how 'the world really is', accompanied by a warning that if we want to change it too much, only totalitarian horror will ensue. Every vision of another world is dismissed as ideology. Alain Badiou put it in a wonderful and precise way: the main function of ideological censorship today is not to crush actual resistance – this is the job of repressive state apparatuses – but to crush hope, immediately to denounce every critical project as opening a path at the end of which is something like a gulag. This is what Tony Blair had in mind when he recently asked: 'Is it possible to define a politics that is what I would call post-ideological?'[24]

In order to understand how ideology functions in its traditional mode, the well-known expression, 'You have to be stupid not to see that!' should be inverted: 'You have to be stupid to see . . .' what? – the supplementary ideological element which provides meaning to a confused situation. In anti-Semitism, for example, you have to be stupid enough to see 'the Jew' as the secret agent who secretly pulls the strings and controls social life. Today, however, the predominant cynical functioning of ideology itself claims: 'You have to be stupid to see that' – what? The hope of radical change.

Notes

INTRODUCTION

1. Alain Badiou, *La vraie vie* (Paris: Fayard, 2016).
2. George Glider, quoted in John L. Casti, *Would-Be Worlds* (New York: John Wiley & Sons, Inc., 1997), p. 215.
3. See Peter Sloterdijk, *Regeln für den Menschenpark* (Frankfurt: Suhrkamp Verlag, 1999).
4. Peter Trawny, *Freedom to Fall: Heidegger's Anarchy* (Cambridge: Polity Press, 2015), p. 98.
5. Martin Heidegger, *Anmerkungen II*, in *Schwarze Hefte 1944–1948*, quoted from Trawny, *Freedom to Fall*, p. 60.
6. The extent to which we are manipulated by our media is easily detected in the gaps in the media's reporting of events. For example, the conflict in eastern Ukraine was for some months treated as a threat to world peace, and then it simply disappeared (from the front pages, at least); it was silently re-normalized. When Ukraine demanded more defensive weapons from the West, it popped up again, and we learned that the fighting has been going on all the time.
7. I owe this information to Zdravko Kobe, Ljubljana.

1. The State of Things

1. See http://yournewswire.com/bill-gates-we-need-socialism-to-save-the-planet/. It is interesting to note that Gates's exact words cannot be independently confirmed.
2. For this view see, among others, vol. 7, no. 1 (March 2017) of *International Critical Thought*, especially the texts by Domenico Losurdo, William Jefferies, Peggy Raphaelle and Cantave Fuyet.
3. Julia Buxton, 'Venezuela After Chavez', *New Left Review* 99, p. 25.

4. Alenka Zupančič, 'Apocalypse, again' (manuscript).
5. In *The Timbuktu School for Nomads* (London: Nicholas Brealey, 2016), Nicholas Jubber provides a detailed description of the daily life of Tuareg nomads in the centre of Saharan Africa, a group of people quite literally 'left behind' by today's rapid globalization. (According to some sources, even their name – Tuareg – originates from 'left behind by god'.) The surprise is that some of the features of their nomadic life appear similar to the nomadic life of the most 'advanced' individuals who are constantly on the move; can we imagine that, provided with contemporary digital machinery (mobile phones, tablets, etc.), Tuaregs would find it easy to immerse themselves in 'postmodern' society with its constant mobility?
6. I owe this point to Karl-Heinz Dellwo.
7. Quoted from https://www.project-syndicate.org/commentary/lesson-of-populist-rule-in-poland-by-slawomir-sierakowski-2017-01.
8. Wolfgang Streeck, *How Will Capitalism End?* (London: Verso Books, 2016), p. 57.
9. See Rebecca Carson, 'Fictitious Capital, Personal Power and Social Reproduction' (manuscript, 2017).
10. Ibid.
11. Ayn Rand, *Atlas Shrugged* (London: Penguin Books, 2007), p. 871.
12. Nina Power, 'Dissing', *Radical Philosophy* 154, p. 55.
13. https://www.thelightphone.com/#lpii
14. David Harvey, personal communication.
15. William James, 'On Some Mental Effects of the Earthquake', quoted from http://storyoftheweek.loa.org/2010/08/on-some-mental-effects-of-earthquake.html.
16. http://www.businessinsider.com/china-social-credit-score-like-black-mirror-2016-10.
17. Quoted from Alfie Bown, *The Playstation Dreamworld* (to appear from Polity Press).
18. http://www.theverge.com/2017/3/27/15077864/elon-musk-neuralink-brain-computer-interface-ai-cyborgs.
19. See Julian Assange, *When Google Met WikiLeaks* (New York: OR Books, 2014).
20. Mike Wehner, 'Scientists remotely hacked a brain, controlling body movements', quoted from http://bgr.com/2017/08/18/brain-hack-science-limb-control/.
21. http://www.cnn.com/2017/10/17/politics/president-donald-trump-alexis-tsipras-greece-evil/index.html.

22. See Jonathan Dickstein and Gautam Basu Thakur, *Lacan and the Posthuman* (London: Palgrave-Macmillan, 2017).
23. I rely in this text on the ideas of many friends of mine, especially Matthew Flisfeder and Todd MacGowan.
24. Todd MacGowan, personal communication.
25. The film just extrapolates the tendency, which is already booming, of more and more perfect silicon dolls. See Bryan Appleyard, 'Falling in Love With Sexbots', *Sunday Times*, 22 October 2017, pp. 24-5: 'Sex robots may soon be here and up to 40 per cent of men are interested in buying one. One-way love may be the only romance of the future.' The reason for the power of this tendency is that it really brings nothing new: it merely actualizes the typical male procedure of reducing the real partner to a support of his fantasy.
26. Matthew Flisfeder, '*Blade Runner 2049* in Perspective' (unpublished manuscript).
27. Ibid.
28. Todd MacGowan, personal communication.
29. I owe this point to Peter Strokin, Moscow.
30. Quoted from https://www.theguardian.com/film/2017/oct/06/blade-runner-2049-dystopian-vision-seen-things-wouldnt-believe.
31. Quoted from https://www.theguardian.com/film/2017/oct/06/blade-runner-2049-dystopian-vision-seen-things-wouldnt-believe.
32. One of the reproaches to my reading was, as expected, that, as a Marxist, I ignore the Gnostic dimension of the film – here is a representative case:

> 1. The names. Joshi is very close to Jeshua (Jesus) and means light bringer, related to the Sun. A solar symbol, associated with Christianity and order. Deckard sounds a lot like Descartes (I read that Philip K. Dick meant that but am not sure). Niander Wallace. Niander = new man. Wallace = foreigner. K. probably derived from Kafka's *Castle* and *Trial*.
>
> 2. Clear religious and Gnostic references. Jewish tradition has the first couple as being Lilith and Samael. It was said their act of mating was somehow dangerous to the well-being of the universe so God separated them, preventing them from ever uniting again. If they ever came back together the well-being of the universe would be threatened again. Now, the Gnostics thought the God of the Christians was actually a blind god they called Samael. Wallace is blind, he has been described as having a 'god complex', he has a dark angel (Luv), and he is obsessed with replicant reproduction. He also wants replicants to replace humans, conquer

the stars and 'storm heaven'. Further, Joshi declares that if replicants can reproduce, that means they have souls, and that would upset everything. (Cosmin Visan at http://thephilosophicalsalon.com/blade-runner-2049-a-view-of-post-human-capitalism/)

Frankly, I don't see a great interest in these points. If Joshi, the ruthless apartheid enforcer, stands for Joshua/Jesus and the Christian order, what kind of Christianity is this? The one of Trump's alt-right supporters? If Deckard and Rachael are a couple in the line of Samael and Lilith, in what sense could the threat brought about by their mating be compared to the threat brought about by the mating of Deckard and Rachael? And so on.

2. Vagaries of Power

1. Quoted from http://www.marxistsfr.org/ebooks/lenin/state-and-revolution.pdf.
2. Ibid.
3. Ibid.
4. Jacques Lacan, 'La Troisième', in *La Cause Freudienne* (Paris, 2011), 79, p. 32.
5. Jean-Claude Milner, 'Back and Forth from Letter to Homophony', in *Problemi International* no. 1 (Ljubljana, 2017), p. 96.
6. Ibid., p. 30.
7. Ibid., pp. 96–7.
8. See http://www.independent.co.uk/news/science/fury-at-dna-pioneers-theory-africans-are-less-intelligent-than-westerners-394898.html
9. Quoted from http://www.marxistsfr.org/ebooks/lenin/state-and-revolution.pdf.
10. Jean-Claude Milner, *Relire la Revolution* (Lagrasse: Verdier 2016), p. 246.
11. Jean-Claude Milner, 'The Prince and the Revolutionary', quoted from http://crisiscritique.org/ccmarch/milner.pdf.
12. Louis Antoine de Saint-Just, 'Rapport sur les factions de l'étranger', in *Œuvres complètes* (Paris: Gallimard 2004), p. 695.
13. As for the relationship between Hegel's *Logic* and Marx's *Capital*, we should not be sentimental and awed by Lenin's statement that anyone who didn't read Hegel's *Logic* cannot understand *Capital*: Lenin himself read Logic but he didn't really understood it (his limit was the category of *Wechselwirkung*), plus he didn't really understand *Capital*. Here one should be precise: what Lenin did not understand was the – let's risk this term – 'transcendental' dimension of Marx's critique

of political economy, the fact that Marx's *Critique of Political Economy* is not just a critical analysis of economics but simultaneously a kind of transcendental form which enables us to articulate the basic contours of the entire social being (inclusive of ideology) in capitalism.

14. Milner, 'The Prince and the Revolutionary'.
15. Quoted from https://www.haujournal.org/index.php/hau/article/view/hau7.2.021/2980.
16. Milner, 'The Prince and the Revolutionary'.
17. Yannis Varoufakis, *Adults in the Room* (London: Penguin, 2017), p. 6.
18. Ernest Mandel, *Trotsky as Alternative* (London: Verso Books, 1995), p. 81.
19. See http://french.about.com/od/grammar/a/negation_form_2.htm.
20. Kojin Karatani, *Transcritique. On Kant and Marx* (Cambridge, MA: MIT Press, 2003), p. 183.
21. https://www.theguardian.com/commentisfree/2017/apr/25/le-pen-far-right-holocaust-revisionist-macron-left.
22. Quoted from http://www.spiegel.de/international/europe/interview-with-french-president-emmanuel-macron-a-1172745-2.html.
23. See https://www.theguardian.com/film/2017/nov/26/susan-sarandon-i-thought-hillary-was-very-dangerous-if-shed-won-wed-be-at-war – and, the *Guardian* being the *Guardian*, immediately followed by a liberal backlash by James Rubin https://www.theguardian.com/us-news/2017/nov/28/susan-sarandon-is-wrong-about-hillary-clinton.
24. See https://www.project-syndicate.org/commentary/poland-hungary-authoritarian-appeasement-by-slawomir-sierakowski-2017-06.
25. Quoted from https://www.marxists.org/reference/archive/mao/selected-works/volume-8/mswv8_34.htm.
26. https://visegradpost.com/en/2017/10/25/viktor-orban-designates-globalization-and-financial-speculators-as-threats-for-identity/.
27. This critique of Orbán triggered a series of conservative reactions which accused me of ignoring all the problems with refugees and simply condoning their free flow to Europe, which will lead to the end of Europe as we know it. My opponents here totally miss the point: I am fully aware of the problems (cultural incompatibility, etc.), which is why I was also viciously attacked by Leftist liberals. I maintain that, while there should be no taboo topics here (one should even be allowed to enquire whether the influx of immigrants is part of some obscure plan), it is a big step from here to the racist anti-Semitic/anti-Muslim plot theory. To paraphrase Lacan yet again, even if one can prove that the influx of refugees is part of a dark plot to destabilize Europe, this in no way justifies the anti-Semitic/anti-Muslim ideology propagated by Orbán and his consorts

in other European countries – this ideology is wrong in itself, it is pathological *per se*, independently of its partial factual accuracy.

28. David Wallace-Wells, 'Uninhabitable Earth', *New York Magazine*, 9 July 2017, available online at http://nymag.com/daily/intelligencer/2017/07/climate-change-earth-too-hot-for-humans.html.
29. See 'Billionaire bunkers: How the 1% are preparing for the apocalypse', available at http://edition.cnn.com/style/article/doomsday-luxury-bunkers/index.html.
30. Bernie Sanders, 'The Republican budget is a gift to billionaires: it's Robin Hood in reverse', available at https://www.theguardian.com/commentisfree/2017/oct/16/republican-budget-gift-billionaires-bernie-sanders.
31. Jamie Peck, available at https://www.theguardian.com/commentisfree/2017/oct/20/womens-convention-bernie-sanders.
32. https://foreignpolicymag.files.wordpress.com/2018/05/cdu.jpg?w=1500&h=1000&crop=0,0,0,0.
33. Quoted from http://yanisvaroufakis.eu/2012/02/14/the-global-minotaur-interviewed-by-naked-capitalism/#more-1753.
34. https://www.theguardian.com/world/2018/may/11/europe-prepares-countermeasures-against-us-iran-sanctions.
35. See https://www.theguardian.com/commentisfree/2018/jun/02/roseanne-barr-working-class-voice-vanishes-tv.
36. Ibid.
37. Ibid.
38. Ibid.
39. Yanis Varoufakis, quoted from https://www.theguardian.com/commentisfree/2018/jun/11/trump-world-order-who-will-stop-him.
40. Milner, *Relire la Revolution*, p. 259.
41. Ibid., pp. 260–61.
42. Quoted from Neil Harding, *Leninism* (Durham: Duke University Press, 1996), p. 309.
43. Ibid., p. 152.
44. Ibid., p. 87.
45. Ibid.
46. Quoted from http://tomclarkblog.blogspot.com/2010/12/curzio-mala parte-bolshevik-coup-detat.html
47. Ibid.
48. Quoted from https://archive.org/stream/CurzioMalaparteTechnique-CoupDEtatTheTechniqueOfRevolution/Currzio+Malaparte++Techni que+Coup+D%27Etat++The+Technique+Of+Revolution_djvu.txt.

3. From Identity to Universality

1. Robert Barnard, *A Talent to Deceive: An Appreciation of Agatha Christie* (London: Fontana Books, 1990), p. 202.
2. Quoted from Bernard Brščič, 'George Soros is one of the most depraved and dangerous people of our time' (in Slovene), *Demokracija*, 25 August 2016, p. 15.
3. Alenka Zupančič (personal communication).
4. Even Trump's decision to recognize Jerusalem as the capital of Israel fits perfectly this logic of Zionist anti-Semitism.
5. Incidentally, since I occasionally contribute comments for the *Russia Today* website, I was put on the list of Putin's 'useful idiots' by a weird body called 'European Values', which is dedicated to 'protecting freedoms'. Life never fails to surprise you: having not only criticized Putin but even applied to him the term 'Putogan' (a conflation of Putin and Erdoğan), and having repeatedly and staunchly defended the emancipatory core of European tradition, I am now Putin's 'useful idiot!'. Well, the only thing we can be sure about is that the guys at European Values are *useless* idiots.
6. See V. I. Lenin, 'The Right of Nations to Self-Determination' (1914), in *Collected Works*, Vol. 20 (Moscow: Progress Publishers, 1972), pp. 303–454.
7. Margaret Washington, on http://www.pbs.org/wgbh/amex/brown/filmmore/reference/interview/washington05.html.
8. See Jamil Khader, 'Against Trump's White Supremacy: Embracing the Enlightenment, Renouncing Anti-Eurocentrism' (quoted from the manuscript).
9. For a scathing critical analysis of liberal PC speech, see Reni Eddo-Lodge, *Why I'm No Longer Talking to White Men About Race* (London: Bloomsbury, 2017).
10. Jamil Khader, 'Against Trump's Supremacy'.
11. Quoted from https://cominsitu.wordpress.com/2017/07/05/the-myth-of-cultural-appropriation/.
12. See Michaele L. Ferguson, 'Neoliberal feminism as political ideology', *Journal of Political Ideologies*, Vol. 22, No 3.
13. Ibid.
14. Ibid.
15. I owe this story and its reading to Alenka Zupančič.
16. See http://www.identitytheory.com/interview-john-summers-baffler/.

17. Quoted from https://www.theguardian.com/society/2018/mar/01/how-americas-identity-politics-went-from-inclusion-to-division.
18. This idea was suggested to me by Mladen Dolar.
19. Quoted from https://www.theguardian.com/world/2018/may/07/viktor-orban-hungary-preserve-christian-culture.
20. See Claude-Lévi Strauss, *Structural Anthropology* 2 (Chicago: University of Chicago Press, 1983).
21. Wang Lixiong and Tsering Shakya, *The Struggle for Tibet* (London: Verso Books, 2009), p. 77.
22. For a perspicuous description of the tension between environmentalists who want to preserve local habitats and the actual people who live in these habitats in Scotland, see James Hunter, *On the Other Side of Sorrow* (Edinburgh: Birlinn, 2014).
23. See Ramesh Srinivasan, *Whose Global Village? Rethinking How Technology Shapes Our World* (New York: New York University Press, 2017).
24. Ibid., p. 209.
25. Ibid., p. 213.
26. Ibid., p. 224.
27. Susan Buck-Morss, *Hegel, Haiti, and Universal History* (Pittsburgh: University of Pittsburgh Press, 2009), p. 151.
28. Ibid., p. 133.
29. Ibid., pp. 138–9.

4. Ernst Lubitsch, Sex and Indirectness

1. See https://www.theguardian.com/lifeandstyle/2018/mar/17/elena-ferrante-even-after-century-of-feminism-cant-be-ourselves.
2. In what follows, I rely on Yuval Kremnitzer's outstanding intervention at the Lubitsch colloquium in Kino Babylon, Berlin, on 28 January 2017.
3. I owe this point to Jela Krečič.
4. Guillermo Arriaga, *21 Grams* (London: Faber and Faber, 2003) p. 107.
5. Quoted from http://news2read.com/lifestyle/1736162/carnality-and-consent-how-to-navigate-sex-in-the-modern-world#.
6. Quoted from http://www.bbc.com/news/world-us-canada-43218355.
7. See https://www.theguardian.com/books/2018/jan/23/germaine-greer-criticises-whingeing-metoo-movement, and https://www.theguardian.com/books/2018/jan/23/germaine-greer-criticises-whingeing-metoo-movement.

8. Robert Pfaller, *Das schmutzige Heilige und die reine Vernunft* (Frankfurt: Fischer Verlag, 2012).
9. Joseph Kerman, *Opera as Drama* (Berkeley: University of California Press, 1988).
10. See pp. 66–7 of the present book.
11. See Moshe Lewin, *Lenin's Last Struggle* (translation of the French original published in 1968) (Ann Arbor: University of Michigan Press, 2005).
12. Ibid., p. 84.
13. Ibid., p. 132.
14. 'Better Few, But Better', available online at www.marxists.org/archive/lenin/works/1923/mar/02.htm.
15. I owe this line of thought to Jela Krečič.
16. Incidentally, the ruthless egotism of Mia and Sebastian is clearly signalled by two incidents. When Mia is late for her date with Sebastian and looks for him in a movie theatre in the middle of the screening, she just steps in front of the screen and shouts his name, indifferent to how this may disturb the audience. When Seb visits her in Boulder, he stops his car in front of her parents' house and loudly honks the horn, not caring about how this might disturb the tired neighbours.
17. In my reading, I rely on Duane Rouselle, https://dingpolitik.wordpress.com/2018/02/17/black-panther-as-empty-container/, Christopher Lebron, http://bostonreview.net/race/christopher-lebron-black-panther, and on an email exchange with Todd McGowan.
18. Paraphrased from https://en.wikipedia.org/wiki/Black_Panther_(film).
19. We leave out of consideration a simple but pertinent question: what kind of country is Wakanda in its political structure? Obviously a monarchy in which the king makes decisions after consulting a close circle of his elite – there is no mechanism for the popular will to be consulted. And what about its economic structure? The film totally ignores this aspect: who controls the (super)natural wealth on which the prosperity of Wakanda is based? Obviously, it's again the king and its clique . . .
20. And this conclusion brings us back to Lenin, who was ready to brutally fight his enemies but who, when the struggle was over, never confused the political conflict with personal animosity and even helped his vanquished enemies – this distance was completely obliterated by Stalin for whom 'the political is personal' was true in a direct obscene sense.

Conclusion: For How Long Can We Act Globally and Think Locally?

1. Quoted from https://www.marxists.org/reference/archive/hegel/works/ae/part2-section3.htm#c3-3-b.
2. Quoted from https://www.marxists.org/reference/archive/hegel/works/hp/hpsocrates.htm.
3. Quoted from https://monoskop.org/images/0/05/Hegel_GWF_Aesthetics_Lectures_on_Fine_Art_Vol_1_1975.pdf.
4. Alenka Zupančič, 'Back to the Future of Europe' (unpublished manuscript).
5. http://www.latimes.com/opinion/op-ed/la-oe-marche-left-fake-news-problem-comedy-20170106-story.html.
6. David Rennie, 'How Soviet sub officer saved world from nuclear conflict', *Daily Telegraph*, 14 October 2002.
7. Quoted from https://www.independent.co.uk/arts-entertainment/tv/news/susan-sarandon-hillary-clinton-america-war-president-donald-trump-feud-bette-and-joan-a8077651.html.
8. Jonathan Brent and Vladimir P. Naumov, *Stalin's Last Crime* (New York: HarperCollins, 2003), p. 307.
9. Ibid., p. 297.
10. See http://abcnews.go.com/International/russian-president-vladimir-putin-unveils-nuclear-weapons-listen/story?id=53435150.
11. Alain Badiou, *Je vous sais si nombreux* . . . (Paris: Fayard, 2017), pp. 56–7.
12. http://www.newsweek.com/egypt-atheism-illegal-crackdown-non-believers-religion-islam-772471.
13. https://www.egypttoday.com/Article/2/40633/OPINION-The-atheists-are-coming.
14. See https://www.theguardian.com/news/2018/mar/17/data-war-whistle blower-christopher-wylie-faceook-nix-bannon-trump.
15. For a precise description of this predicament see Laurent de Sutter, *Narcocapitalism* (Cambridge: Polity Press, 2018).
16. Quoted from http://th-rough.eu/writers/bifo-eng/journey-seoul-1.
17. Frank Ruda, personal communication. All non-attributed quotes in this chapter are from the same source.
18. Robert Pippin, 'Hegel on the Varieties of Social Subjectivity', in *German Idealism Today*, ed. Markus Gabriel and Anders Moe Rasmussen (Boston: De Gruyter, 2017), pp. 132–3.

19. In *America Day by Day*, quoted from Stella Sandford, *How to Read Beauvoir* (London: Granta Books, 2006), p. 42.
20. Ibid., p. 49.
21. https://www.theguardian.com/world/2017/dec/08/lead-us-not-into-mistranslation-pope-wants-lords-prayer-changed.
22. Soren Kierkegaard, *Fear and Trembling* (Princeton: Princeton University Press, 1983), p. 115.
23. Pippin, 'Hegel on the Varieties of Social Subjectivity', pp. 134–5.
24. Quoted from http://www.newyorker.com/culture/persons-of-interest/the-return-of-tony-blair.

SEVEN STORIES PRESS is an independent book publisher based in New York City. We publish works of the imagination by such writers as Nelson Algren, Russell Banks, Octavia E. Butler, Ani DiFranco, Assia Djebar, Ariel Dorfman, Coco Fusco, Barry Gifford, Martha Long, Luis Negrón, Peter Plate, Hwang Sok-yong, Lee Stringer, and Kurt Vonnegut, to name a few, together with political titles by voices of conscience, including Subhankar Banerjee, the Boston Women's Health Collective, Noam Chomsky, Angela Y. Davis, Human Rights Watch, Derrick Jensen, Ralph Nader, Loretta Napoleoni, Gary Null, Greg Palast, Project Censored, Barbara Seaman, Alice Walker, Gary Webb, and Howard Zinn, among many others. Seven Stories Press believes publishers have a special responsibility to defend free speech and human rights, and to celebrate the gifts of the human imagination, wherever we can. In 2012 we launched Triangle Square books for young readers with strong social justice and narrative components, telling personal stories of courage and commitment. For additional information, visit www.sevenstories.com.